The Atlas of British Railway History

First published in 1985, this Atlas uses over 50 specially drawn maps to trace the rise and fall of the railways' fortunes, and is supported by an interesting and authoritative text. Financial and operating statistics are clearly presented in diagrammatic form and provide a wealth of information rarely available to the student of railway history. Freeman and Aldcroft provide the basis for a new understanding of the way in which the railways transformed Britain by the scale of their engineering works, by shrinking national space and reorganising the layouts of urban areas. Maps show the evolution of early wagon routes into the first railway routes, the frenetic activity of the 'Railway Mania' years, and the consolidation of these lines into a national network. This exciting presentation of railway development will interest the enthusiast as well as the more general student of British transport history.

The Atlas of British Railway History

Michael Freeman and Derek Aldcroft

Routledge
Taylor & Francis Group

First published in 1985
by Croom Helm

This edition first published in 2018 by Routledge
2 Park Square, Milton Park, Abingdon, Oxon, OX14 4RN

and by Routledge
711 Third Avenue, New York, NY 10017

Routledge is an imprint of the Taylor & Francis Group, an informa business

Publisher's Note
The publisher has gone to great lengths to ensure the quality of this reprint but points out that some imperfections in the original copies may be apparent.

Disclaimer
The publisher has made every effort to trace copyright holders and welcomes correspondence from those they have been unable to contact.
A Library of Congress record exists under ISBN: 0709905424

ISBN 13: 978-1-138-56633-0 (hbk)
ISBN 13: 978-1-315-12376-9 (ebk)

The Atlas of
BRITISH RAILWAY
HISTORY

Michael Freeman and Derek Aldcroft

CROOM HELM
LONDON. SYDNEY. DOVER, NEW HAMPSHIRE

© 1985 Michael Freeman
Croom Helm Ltd, Provident House, Burrell Row,
Beckenham, Kent BR3 1AT
Croom Helm Australia Pty Ltd, Suite 4, 6th Floor,
64-76 Kippax Street, Surry Hills, NSW 2010, Australia

British Library Cataloguing in Publication Data

Freeman, Michael
   The atlas of British railway history.
   1. Railroads—Great Britain—History
   I. Title   II. Aldcroft, Derek H.
   385'.0941   HE3018
     ISBN 0-7099-0542-4

Croom Helm, 51 Washington Street, Dover,
New Hampshire 03820, USA

Library of Congress Cataloging in Publication Data

Freeman, Michael, 1950–
   The atlas of British railway history.
   1. Railroads—Great Britain—History
   I. Aldcroft, Derek Howard.   II. Title.
   TF57.F84   1985   385'.0941   85-16645
   ISBN 0-7099-0542-4

Typeset by Words & Pictures Ltd, Thornton Heath, Surrey

CONTENTS

The justification for compiling an atlas of British railway history may appear obscure to some eyes. Naturally one can hide behind the familiar claim that nothing quite like it has been attempted before. The publication of new photographs further supports such an argument. However, it is not difficult to find reasons for an overtly geographical perspective. A fundamental characteristic of British railway development in the nineteenth century, for example, was the geographically fragmented nature of control. There was no real sense in which one could talk of a *British* railway system. Railway operation was divided between a multiplicity of independent companies, each with its distinctive territorial identity and each with its own way of doing things — from operations to architecture. Equally significant was the measure in which railways transformed the country's geography. Railways were great 'connectors', notionally drawing towns and cities together, and in total having the effect of shrinking the geographical size of the country: the substitution of machine for muscle transformed the frictional effect of distance on movement. Railways also had demonstrable impact upon the shape of cities and towns. Rail lines out of London became foci for the spreading tentacles of suburbia. In the centres of cities, railways came to be a prime determinant of the layout of streets and buildings, not to mention their own occupancy of vast areas of central land. A yet more simple geographical view lies in the record of the spread and contraction of Britain's railway system. Some parts of the country gained railway communication long before others.

Some benefited from being served by several different companies. The network maps in this book reveal an interesting symmetry in the sequence of openings and closures, the last lines opened often being the first to be closed, with the modern-day network showing an uncanny resemblance to that of 1850.

The economic significance of British railways, in particular their role as causal agents of Victorian economic growth, is the subject of some dispute amongst historians. But there can be no doubt of the railways forming a primary economic sector throughout the Victorian age, nor of their symbolic importance in society at large. It is conventional to see the railways as the force which bound region and province into nation, as the essential medium of cultural convergence. But this must not lead one to overlook the railways' function as validators of local resource wealth and, in consequence, as part creators of the regions of vigorous industrial expansion upon which the Victorian age so depended. Herein rests an important paradox, one that is very apparent in the pages that follow. It was partly for this reason that the amalgamation of companies reached a point of stalemate after 1870.

As the twentieth century draws to a close, it is increasingly likely that generations will grow up for whom railways are a specialised and geographically restricted means of inland transport. This volume reveals how once upon a time the railways were the supreme common carrier, their influence extending to almost every aspect of culture and economy.

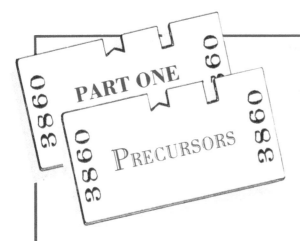

PART ONE

PRECURSORS

3860 3860 3860 3860

The railway principle is not a modern one. The Greeks used 'rutways' to reduce tractive effort. The Romans laid down flat blocks of stone to reduce friction on chariot wheels. However, the truest precursors of the modern railway were the mining waggonways of central Europe, laid from the sixteenth century. The first British waggonway probably came in to use around 1630 to serve collieries near Newcastle. On the early systems rails were made of timber and wore down easily. Subsequently, therefore, iron plates were fastened to them, giving the term plateway.

Plate 1: A laden coal wagon descending to a riverside coal staith.

Later still, the plate rails were replaced with
cast iron ones; the great Coalbrookdale
Ironworks made its first cast iron rails in 1767.
At first, waggons were prevented from slipping
off the tracks by incorporating a flange into the
rails, but in 1789 a feeder railway to the
Loughborough Canal transferred the flange to
the inside of the wheel and this provided the
model upon which construction is based today.
The map below depicts part of a waggonway system
which extended inland for a total of sixteen miles
and southward to the River Wear for another
ten miles.

Waggonways on the lower Tyne in Northumberland in the eighteenth century, with dates of opening and closure where known.

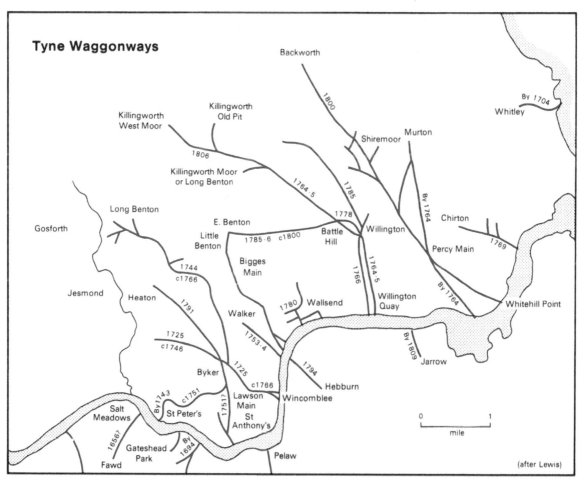

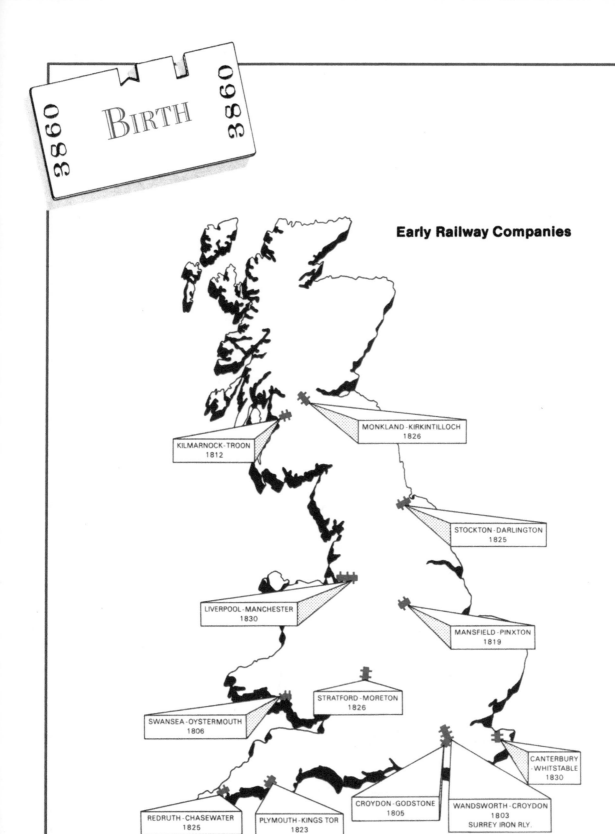

Early Railway Companies

MONKLAND - KIRKINTILLOCH
1826

KILMARNOCK - TROON
1812

STOCKTON - DARLINGTON
1825

LIVERPOOL - MANCHESTER
1830

MANSFIELD - PINXTON
1819

STRATFORD - MORETON
1826

SWANSEA - OYSTERMOUTH
1806

CANTERBURY
- WHITSTABLE
1830

REDRUTH - CHASEWATER
1825

PLYMOUTH - KINGS TOR
1823

CROYDON - GODSTONE
1805

WANDSWORTH - CROYDON
1803
SURREY IRON RLY.

*Plate 2: One of the locomotives built by
William Hedley for use on the Wylam Colliery
Railway, 1813.*

By 1800 there were many hundreds of miles of wooden waggonways and plateways, especially in the extensive mining districts of the North East, the Midlands and South Wales. But they attracted little attention from the public at large. There were no lines carrying passengers, while the colliery systems, for all their numerical significance, were essentially piecemeal in conception, ephemeral in existence and primitive in engineering. The watershed was reached in 1801 when Parliament passed an Act enabling the Surrey Iron Railway Company to construct part of a line from London to Portsmouth. Although the technology was still primitive, the scheme established the first railway company and the first public railway. It brought the railway principle before the eyes of legislators and, by the proposal to link London with the naval arsenal of Portsmouth, widened ideas about the railway's potential geographical significance. Between 1801 and 1821 some nineteen further Acts were passed. Among these, the Swansea and Oystermouth Railway provides the first authenticated instance of the carriage of passengers. The first railway companies were evenly spread throughout the country. Some had grown from a tradition of waggonway-working, but others had no such connection — witness to the railway principle's widening currency.

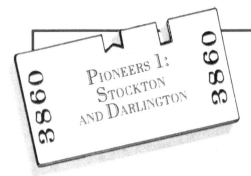

PIONEERS 1:
STOCKTON
AND DARLINGTON

3860
3860

To even the most disinterested of laymen, the Stockton and Darlington Railway is well known as a landmark in railway history. It was not the first public railway. But it was the first public railway regularly to employ steam locomotives to haul wagons of goods and carriages of passengers. The railway was conceived as a more efficient means of carrying coal and other minerals from Darlington and its vicinity to Stockton, the River Tees and thence to the sea. The gestation period was exceedingly protracted. Plans were first mooted in 1810 and for nearly ten years there was an argument whether a canal would be better than a railway. Eventually an Act was obtained in 1821. It comprised 67 closely printed pages, incorporating the whole of the law relating to railways. In its original form the Act provided for haulage by men or horses. It was thus little different from its immediate predecessors. Shortly after this, however, George Stephenson became involved in the scheme and, through his influence, the Act was amended to allow the use of stationary and moveable steam engines, the former for installation on two inclined planes. The railway opened to traffic in September 1825 and was an immediate success. On the first day it was shown that, on an incline, one engine could draw a train of 80 tons at ten to fifteen miles per hour. As anticipated by its promoters, coal formed the primary traffic. There was a limited conveyance of passengers, arranged not by the railway company but by local road carriers.

Smith's map of the County of Durham (1827) showing the line of the Stockton and Darlington Railway and its various branches.

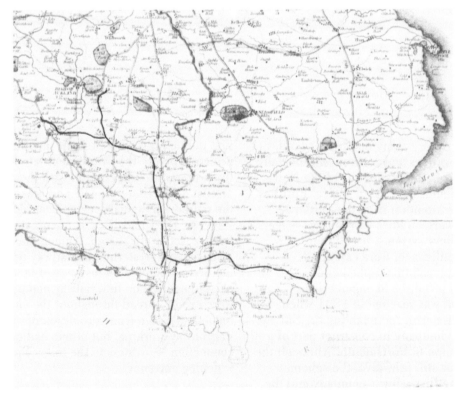

Plate 3: George Stephenson's Locomotion, *built for the Stockton and Darlington Railway in 1825.*

Plate 4: Horse-drawn passenger carriage on the Stockton and Darlington Railway, 1825. As yet the steam locomotive had no monopoly over traction.

While the Stockton and Darlington Railway was coming into being, another railway venture was being planned which later stole the limelight and laid counter-claim to marking the beginning of the railway age. The Liverpool and Manchester Railway was conceived in the early 1820s as a result of rising dissatisfaction with existing road and water conveyance. Monopolistic practices among waterway proprietors, which raised freight charges to exorbitant levels, were a key element. But there were also growing logistical problems for trade between the two towns. The cotton industry was by 1820 growing on such a scale that, whatever the charges, existing transport capacity was hopelessly inadequate.

A company for the railway's promotion was formed in 1824 and after much opposition from local landowners and within Parliament an Act was obtained in 1826. As an engineering project the line faced some formidable difficulties, including the making of a cutting 70 feet deep and nearly two miles long at Olive Mount, and the crossing of the immense bog called Chat Moss. Construction was almost complete by the end of 1828, but a decision remained to be made on the motive power to be employed. Initial ideas had favoured cable haulage operated by stationary engines. Later, the directors were persuaded to employ locomotive engines and instituted the now famous Rainhill contest to find the most suitable design. George Stephenson, already resident engineer on the line, won the contest with the *Rocket* and seven further Stephenson locomotives were ordered for the formal opening on 15th September, 1830. The Liverpool and Manchester lays claim to being the prototype of the modern railway because it used locomotive engines throughout its length, was the first to take seriously the carriage of passengers and was motivated by public welfare as well as private profit.

Plate 5: Liverpool Edge Hill, showing the tunnels down to Wapping and the cables used for drawing the trains.

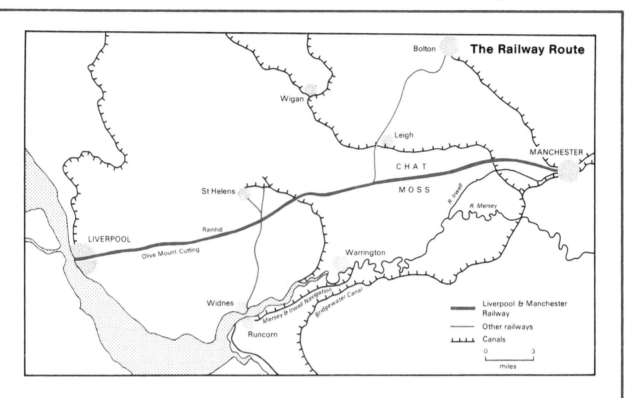

Plate 6: Sankey Viaduct on the Liverpool and Manchester Railway.

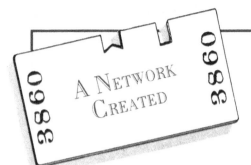

The Railway Map Circa 1840

A Grand Junction
B London & Birmingham
C Great Western
D Bristol & Exeter
E London & Southampton
F London & Brighton
G Birmingham & Gloucester
H North Midland
J Hull & Selby
K Great North of England
L Newcastle & Carlisle

0   50
miles

The Liverpool and Manchester met with a level of success which even its most optimistic promoters had never dared forecast. There were teething troubles in operation, but financially the railway proved a gold mine. Within two years of opening, the company's shares had doubled in value. Passenger traffic on the line had exceeded all expectations; the railway had seemingly created a new travelling public. It was not long, therefore, before clamour for railways became a national pastime. Liverpool soon turned its eyes towards a link with Birmingham; and that city,

Plate 7: The monumental portico of Box Tunnel, near Bath, on Brunel's Great Western Railway.

in turn, looked towards London. Thus were created the Grand Junction Railway and the London and Birmingham Railway, the former representing the first trunk railway in Britain when opened between Warrington and Birmingham in 1837. Further major lines followed, including the Newcastle and Carlisle in 1839, the London and Southampton in 1840 and the pioneering Great Western scheme from London to Bristol in 1841.

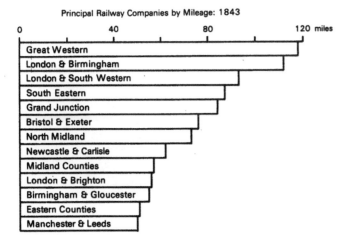

Principal Railway Companies by Mileage: 1843

- Great Western
- London & Birmingham
- London & South Western
- South Eastern
- Grand Junction
- Bristol & Exeter
- North Midland
- Newcastle & Carlisle
- Midland Counties
- London & Brighton
- Birmingham & Gloucester
- Eastern Counties
- Manchester & Leeds

THE RECORD OF CONSTRUCTION COSTS TO 1843 (£s per mile)

DEAREST LINES			CHEAPEST LINES		
Manchester and Leeds	(50m)	£62,080	Whitby and Pickering	(24m)	£5,625
Manchester and Birmingham	(31)	£60,967	Stockton and Darlington	(38)	£6,736
Great Western	(118)	£56,253	Leeds and Selby	(20)	£17,000
Eastern Counties	(51)	£53,682	Newcastle and Carlisle	(61)	£17,393
London and Birmingham	(112)	£53,042	Taff Vale	(30)	£17,966
Liverpool and Manchester	(31)	£49,268	Hull and Selby	(30)	£21,533

Britain's irregular topography and complex geology presented the railway builders with daunting problems. They overcame them through an intriguing combination of laborious trial and error and civil engineering genius, although one unfortunate consequence was the great difficulty of estimating the costs of construction in advance — most schemes exceeded their projected costs, some by vast margins. The relatively easy terrain of the Hull and Selby line understandably made it among the cheapest to build. The most expensive was, predictably, the first line across the Pennines — the Manchester and Leeds. The variation in construction costs cannot be entirely accounted for in this way, though. The Great Western, for instance, was expensive partly because its engineer, I.K. Brunel, insisted on laying out as level a line as possible. Other engineers were content to follow an undulating principle, grading the line within the limits set by the prevailing technology of railway motive power. Brunel's legacy remains with us today: the main line from Paddington to Didcot counts among the best for modern high-speed running.

Plate 8: Construction in progress on the London and Birmingham Railway: the Wolverton embankment.

Plate 9: The approach cutting to Long Tunnel,
Fox's Wood, Great Western Railway.

The Dividend Earners in 1842

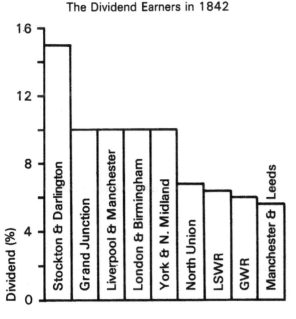

Most of the early main line projects came into
being through the inadequacies of pre-existing
transport media. The complaints of the
Liverpool and Manchester merchants about
congestion and monopolistic pricing were
repeated over and over again as the main line
companies drew up their prospectuses and
charted their courses through Parliament.
However, there were some major projects and
many minor ones which came to fruition
largely through speculation. The first
companies had shown that railways could
generate great profits. And since there were
few regulations governing railway promotion, it
was open to anyone to float a new company
and attract the hungry bands of investors eager
to make their fortunes. The outcome was
'railway mania'.

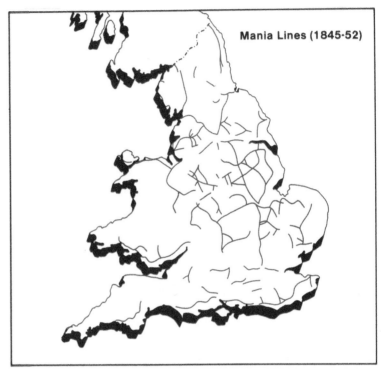

Mania Lines (1845-52)

The several railway manias conferred upon Britain a rapidity of railway development which even a government would have found difficulty in achieving. The first and lesser of the manias occupied the years 1835 to 1837. During that time Acts were passed for 50 new lines totalling about 1,600 miles. Financial backing was forthcoming for the most unrealistic schemes. Potential shareholders were required to provide a deposit of only five per cent and in this way even comparatively small investors could gain a railway interest. Inevitably boom was followed by slump as actual building costs surged ahead of estimates and as Parliament sought to tighten the legislative requirements. The years 1838 and 1839 produced lines totalling but 92 miles; 1840 produced none at all.

The principal railway mania occupied the years 1844 to 1846. Its intensity was startling, as the accompanying charts reveal. In one year alone (1846), 4,540 miles of line were sanctioned, nearly twice the length of the railway system as it then existed. Capital authorisations broke similar records. Many factors accounted for the upsurge of activity.

A run of good harvests, especially favourable loan terms from the Bank of England, and a relaxation of some of the restrictions imposed by Parliament in 1837 were three of the more important. But once any speculative boom has begun it acquires a momentum of its own. The mania brought with it an army of new professions and trades. Railway newspapers were coming into existence almost every week. Bands of 'traffic-takers' ranged the country in search of likely routes. A novel breed was added to the aristocracy in railway magnates like George Hudson. And cartoonists found in the mania a never-ending source of copy. Naturally the tide of speculation waned. Once the shaky foundations of the new capital market were exposed, a precipitate slump occurred, hastened by worsening economic conditions generally. Railway share prices halved between 1846 and 1849. Mileage sanctioned fell to 17 in 1849.

It is easy to deride the railway manias and the lack of proper government control. But the railway system was undeniably advanced ahead of its time, as the following section shows.

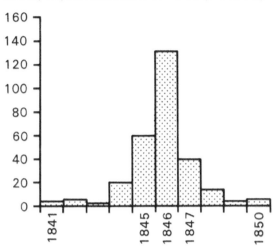

Railway Capital Authorised: 1841-50 (£ millions)

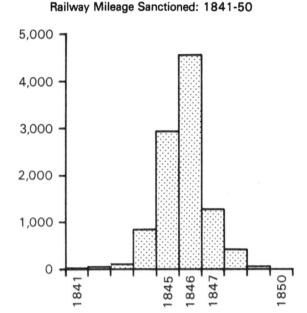

Railway Mileage Sanctioned: 1841-50

PUNCH'S RAILWAY PROSPECTUS.

GREAT NORTH POLE RAILWAY,

Forming a junction with the Equinoctial line, with a branch to the horizon.

Capital, two hundred millions. Deposit, threepence.

DIRECTORS AT THE NORTH POLE.

Jack Frost, Esq., Chairman of the North-west Passage.
Baron Iceberg, Keeper of the Great Seal on the Northern Ocean.

DIRECTOR AT THE HORIZON.

Hugh de Rainbow, Admiral of the Red, Blue, and Orange, &c. &c.

DIRECTORS IN LONDON.

Simon Scamp, Esq., Chairman of the East Jericho Junction Railway.
Thomas Trapper, Esq., Director of the General Aerial Navigation Company.
Sir Edward Alias, Non-Resident-Director of the Equitable Coal and Slate Association.

(With power to add to their number, by taking in as many as possible.)

The proposed line will take the horizon for its point of departure, and, passing over the equator, will terminate at the North Pole, which will be the principal station of the Company.

It is calculated that sunbeams may be conveyed along the line by a new process, which Professor Twaddle has been employed by the provisional committee to discover ; and the professor's report will be laid before the subscribers at the very earliest opportunity.

By bringing the equator within a week of the North Pole, and co-operating with the proprietors of the Great Equinoctial Line, the advantages to the shareholders will be so obvious, that it is hardly necessary to allude to them.

It is calculated that the mere luggage traffic, in bringing up ice from the North Pole to the London market, will return a profit of 65 per cent. on the capital.

Should any unforeseen circumstance occur to prevent the Railway being carried out, the deposit will be returned, on application to Messrs. Walker, Gammon, and Co., Solicitors to the Company, at their temporary offices, in Leg Alley.

THE RAILWAY JUGGERNAUT OF 1845.

Plate 10: The railway mania as depicted in a Punch cartoon.

Plate 11: Satire on a railway prospectus, from Punch, 1845.

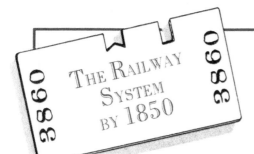

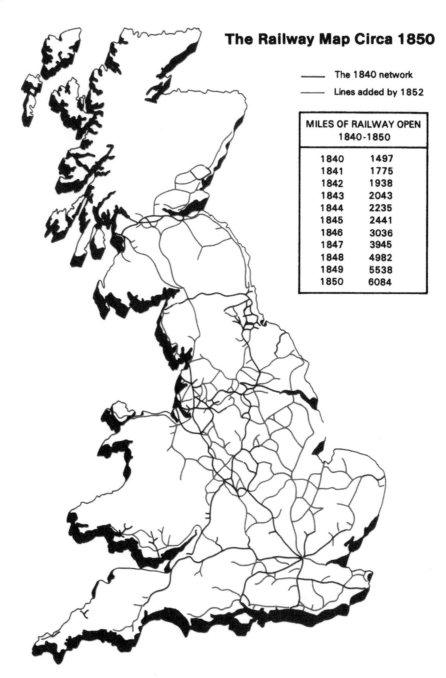

The Railway Map Circa 1850

——— The 1840 network

——— Lines added by 1852

MILES OF RAILWAY OPEN 1840-1850	
1840	1497
1841	1775
1842	1938
1843	2043
1844	2235
1845	2441
1846	3036
1847	3945
1848	4982
1849	5538
1850	6084

The decade between 1840 and 1850 saw a
transformation of the railway map in Britain.
In 1840 there were no more than 1,500 miles
of railway in existence; by 1850 the total
exceeded 6,000. Had all the railways authorised
during the mania been constructed, the figure
would have been nearer 12,000 miles. To all
intents and purposes, Britain could be said to
have possessed a complete railway system
by 1850. The main body of the country was
criss-crossed with trunk and secondary lines.
Only south-west England, Wales and the north
of Scotland had escaped coverage. Limited
traffic prospects and inhospitable terrain made
these areas distinctly uninviting for railway
promotion. The physical extension of railway
communication on such a gargantuan scale and
in so short a time involved a massive
consumption of capital. As such, the railways
brought about a revolution in capital formation
in Britain. In the mania year of 1847, gross
railway capital formation had risen to 6.7 per
cent of national income. This was equivalent to
two-thirds of the value of all domestic exports.
Railway shares became a new and central
ingredient of investment, giving rise to the
growth of provincial stock exchanges. The
public utility, of which rail communication was
a prime example, became a highly preferred
focus of investment for businessmen's profits.

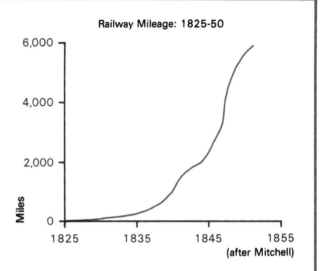

Railway Mileage: 1825-50

(after Mitchell)

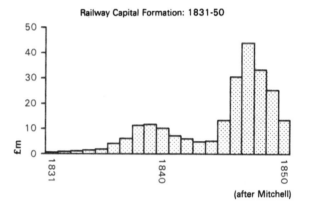

Railway Capital Formation: 1831-50

(after Mitchell)

*Plate 12: The bridge over the River Wye at
Chepstow (GWR).*

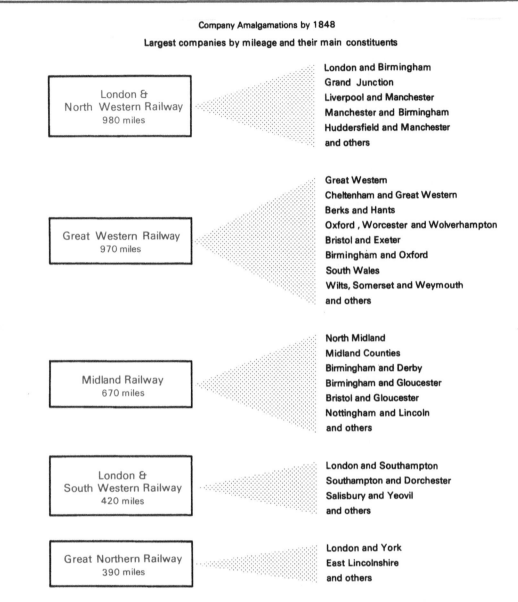

Company Amalgamations by 1848

Largest companies by mileage and their main constituents

London & North Western Railway 980 miles
- London and Birmingham
- Grand Junction
- Liverpool and Manchester
- Manchester and Birmingham
- Huddersfield and Manchester
- and others

Great Western Railway 970 miles
- Great Western
- Cheltenham and Great Western
- Berks and Hants
- Oxford, Worcester and Wolverhampton
- Bristol and Exeter
- Birmingham and Oxford
- South Wales
- Wilts, Somerset and Weymouth
- and others

Midland Railway 670 miles
- North Midland
- Midland Counties
- Birmingham and Derby
- Birmingham and Gloucester
- Bristol and Gloucester
- Nottingham and Lincoln
- and others

London & South Western Railway 420 miles
- London and Southampton
- Southampton and Dorchester
- Salisbury and Yeovil
- and others

Great Northern Railway 390 miles
- London and York
- East Lincolnshire
- and others

If the general form of Britain's railway system by 1850 had been profoundly influenced by speculative investment, it is equally true to say that the precise configuration of lines was as much a by-product of inter-company rivalry as of simple commercial forces. By the close of 1844, Britain had some 100 separate railway companies, and roughly another 100 were to be added by 1850. Early ventures like the Liverpool and Manchester faced no competition from rival railway enterprises. But as numbers of lines and companies multiplied, so competition grew. Control of trunk routes was the main battleground. The resulting Parliamentary and boardroom contests have been reviewed as no less than a form of imperialism which left few parts of the country untouched. The leading protagonists soon realised that their companies' interests here would be better served by absorbing contiguous but lesser railway enterprises. Thus began the great movement towards amalgamation and territorial consolidation. By 1848 most of the companies which were to dominate the British railway scene later in the century had come into being.

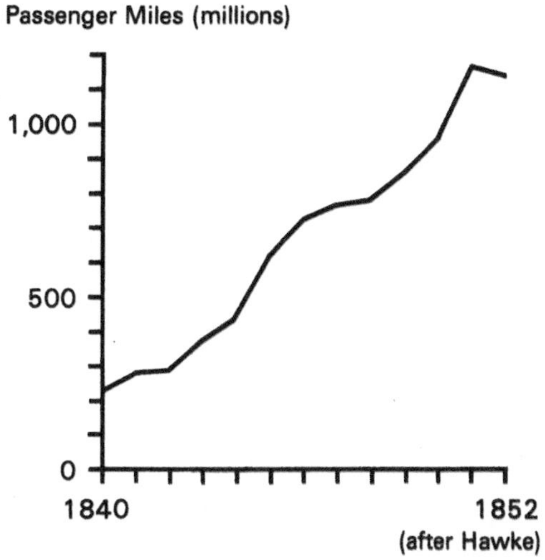
Passenger Miles (millions)
1,000
500
0
1840
1852
(after Hawke)

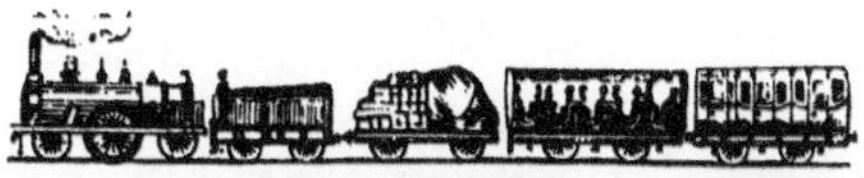

CALEDONIAN RAILWAY TIME TABLE,

ON AND AFTER 10TH SEPTEMBER 1847.

(Until further Notice).

CARLISLE AND BEATTOCK.

NOTICE. – *The Doors of the Booking Office will be Closed punctually at the Hours fixed for the Departure of the Trains, after which no Person can be admitted – Passengers, to ensure being booked, should arrive at the Stations and obtain their Tickets Ten Minutes earlier than the Times mentioned in the following Table.*

Distance.	Edinburgh and Glasgow to Carlisle, &c.	CLASS 1 2 & 3	CLASS 1 & 2	CLASS 1 & 2
			A.M.	P.M.
	EDINBURGH by Coach.......	...	7 0	4 30
	GLASGOW do.	...	5 15	2 30
Miles.	Trains leave	A.M.	P.M.	
5	BEATTOCK.....................	6 30	1 0	10 30
10¾	WAMPHRAY.....................	6 45	1 15	......
13¾	NETHERCLEUCH.....................	7 5	1 33	.....
19¼	LOCKERBIE.....................	7 15	1 45	11 15
26½	ECCLEFECHAN.....................	7 30	2 0	11 30
31	KIRKPATRICK.....................	7 50	2 20	......
35½	GRETNA.....................	8 5	2 35	......
39½	ROCKLIFF.....................	8 18	2 48	......
	Arrive at CARLISLE............	8 30	3 0	12 30
		P.M.		A.M.
	PRESTON.....................	1 25	7 53	5 10
	MANCHESTER.....................	3 5	.. ...	6 35
	LIVERPOOL.....................	3 15	9 45	6 45
			A.M.	P.M.
	LONDON.....................	......	5 32	1 0

Distance.	Carlisle, &c. to Edinburgh and Glasgow.	CLASS 1 & 2	CLASS 1 2 & 3	CLASS 1 & 2
	Trains leave	P.M.	A.M.	A.M.
	LONDON.....................	8 45	......	10 0
	LIVERPOOL.....................	......	7 30	3 50
	MANCHESTER.....................	.....	7 40	4 0
	PRESTON.....................	5 35	9 35	5 35
		A.M.	P.M.	P.M.
Miles.	CARLISLE.....................	10 0	2 30	10 10
4	ROCKLIFF.....................	10 12	2 42	.....
8½	GRETNA.....................	10 25	2 55	10 45
13	KIRKPATRICK.....................	10 40	3 10	......
20	ECCLEFECHAN.....................	11 0	3 30	10 10
25¾	LOCKERBIE.....................	11 15	3 45	11 25
28¾	NETHERCLEUCH.....................	11 25	3 55	... ..
34½	WAMPHRAY.....................	11 47	4 15	......
39½	Arrive at BEATTOCK............	12 0	4 30	12 ·0
	EDINBURGH by Coach....	6 P.M.		6 A.M.
	GLASGOW by Coach............	7 43		7 33

FARES.

	First Class and Inside Coach.	Second Class and Outside Coach.	
Between GLASGOW AND CARLISLE,........................	30/6	20/-	
... EDINBURGH AND CARLISLE,........................	29/6	19/6	
... CARLISLE AND BEATTOCK,........................	8/6 First Class.	5/6 Second Class.	3/4 Third Class.

_ The Company will not be answerable for any Luggage, unless Booked and Paid for; and, for better security, Passengers are recommended to take Carpet Bags and small Packages inside the Carriages, and to have their Address written on all their Luggage in full.—Children under Ten Years of Age, Half Price; Children in arms, unable to walk, pass Free.

HORSES.—Grooms in charge of Horses to pay Second Class Fares.—The Company will not be liable in any case for loss or damage to any Horse or other Animal above the value of £40, unless a declaration of its value, signed by the Owner or his Agent at the time of booking, shall have been given to them, and by such declaration the Owner shall be bound; the Company not being in any event liable to any greater amount than the value so declared. The Company will in no case be liable for injury to any Horse or other Animal, of whatever value, where such injury arises wholly or partially from fear of restiveness.—If the declared value of any Horse or other animal exceed £40, the price of conveying will, in addition to the regular fare be after the rate of 2½ per cent., or 6*d.* per pound upon the declared value above £40, whatever may be the amount of such value, and for whatever distance the Horse or other Animal is to be carried.

CARRIAGES.—Passengers travelling by the Railway with Private Carriages are charged First Class, and their Servants Second Class Fares; and corresponding Tickets are issued for each Class, which are also available for the Company's Carriages. To prevent mistakes, Passengers are requested to declare in each case the number of Servants to whom Second Class Tickets are to be issued.

N.B.—The Servants of the Company are prohibited from demanding or receiving any Gratuity from Passengers, who, it is hoped, will assist the Directors in enforcing this Regulation. Immediate dismissal follows the discovery of any Servant of the Company receiving any Gratuity.

_ *Smoking in the Carriages and at the Stations is forbidden, under a Penalty, by Act of Parliament.*

Trains marked thus (*) do not run on Sundays.

Booking Office in EDINBURGH, Nos. 2 & 10, Princes Street.

... ... GLASGOW, Walker & Co., Tontine Hotel, and A. Mein & Co., Trongate.

By Order,

J. W. CODDINGTON, *Secretary.*

PETER BROWN, PRINTER, EDINBURGH.

THE BROAD GAUGE

Plate 13: A broad gauge train on the Great Western Railway. The mixed gauge track allowed narrow gauge trains to work over the same line.

The standard track gauge of 4ft 8½in. found on Britain's railways today did not enjoy equivalent supremacy in the mid-nineteenth century. The Great Western Railway's renowned engineer, I.K. Brunel, employed a track width of 7ft 0¼in. as he could find no engineering rationale for adopting the smaller gauge. When the speeds proposed on the Great Western were considered and the masses to be moved, Brunel concluded that an altogether larger scale was required. However, as Britain's railway system grew, the operational problems arising from having different gauge networks became manifest. Parliament ordered an inquiry and the result was the Gauge Act of 1846 which defined 4ft 8½in. as 'standard' unless special statutory provision was made. It was a defeat for Brunel, but it did not bar the GWR and its associated companies from retaining the broad gauge, nor from expanding it. Not until the 1860s did the GWR management finally abandon their practice. Already there were many miles of broad gauge

track laid for mixed gauge working, but thereafter wholesale conversion to standard gauge became the task. The last broad gauge train left Paddington in May 1892.

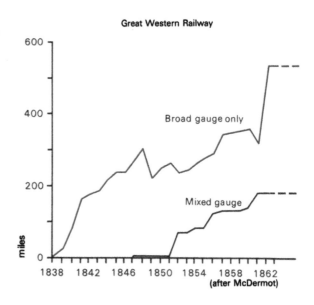

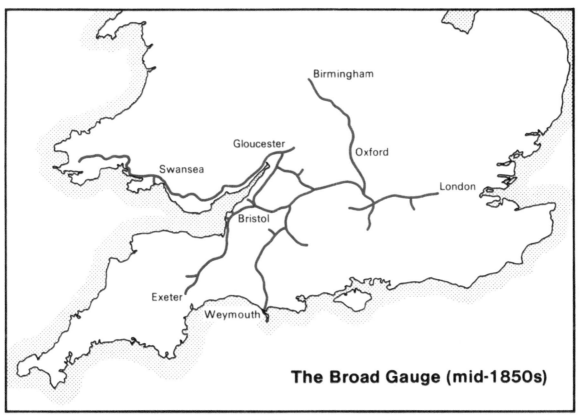

The Broad Gauge (mid-1850s)

Some of the lines shown on the map were also laid for mixed gauge working, for instance the section from Oxford to Birmingham.

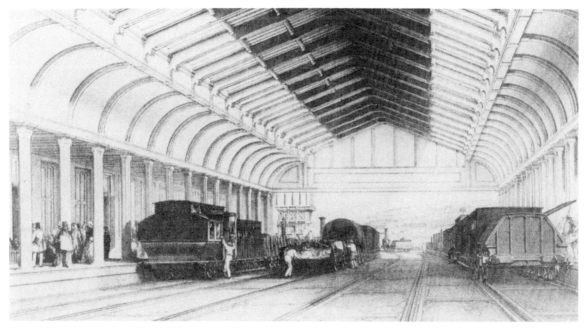

Plate 14: Bath station, Great Western Railway, showing broad gauge tracks and trains.

Railways brought about a revolution in the speed of travelling. The substitution of the mechanical power of steam for the muscle power of horses allowed a transformation in the rate of motion. The principal stage coaches on Britain's turnpike roads had realised the technical limits of their speed by at least 1830. By that time average rates of travelling were pitched around 10 miles per hour. A journey from London to Norwich took 12 hours. Liverpool, Manchester and the Yorkshire towns were some 30 hours away. By 1845 railways had effectively shrunk the country's geographical space to between one-third and one-fifth of its former scale. Express trains reached Norwich in 4½ hours, Liverpool in a little under 7 hours and York in 9 hours. Speeds varied between companies, with the GWR providing some of the best on Brunel's

broad gauge lines westward from London. By August 1845 it was possible to travel from Paddington to Bristol in just 2 hours and 40 minutes. In coaching days the journey had taken the best part of 20 hours. The early railways were not paragons of efficiency, however. Bad time-keeping consequent upon mechanical failures, mishaps and deficient operations management was a persistent feature and caused much public complaint. Some companies became notorious for this, as they often did also for their slow speeds. The Great Western generally won the greatest praise, an undoubted tribute to the superiority of its engineering and management.

Plate 15: The Wootton Basset incline on Brunel's London-Bristol main line.

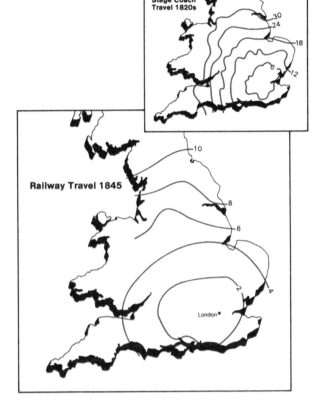

Comparative travel times from London by stage coach and by railway (six-hourly and two-hourly isochrones respectively)

Plate 16: The Great Hall at Euston, London and Birmingham Railway. The architect was Philip Hardwick. This structure, along with the famous Doric portico, was demolished to make way for a new station in the 1960s.

The Cost of Travel

The map shows how far one could travel for one pound (20 shillings) at the time of the transition from road coach to railway in the 1840s.

Third-class rail

Second-class rail

Outside stage coach

LONDON

Passenger travel before the railways was always expensive. Stage coach fares of 3½ to 4½d (1½p) a mile for an inside seat and 2d to 2½d (1p) a mile outside were beyond the reach of the labouring classes with their daily earnings of little above two shillings (10p). The much greater working efficiency of the steam-hauled train against the horse-drawn coach at last made it possible to cater for the transportation of the masses. And, ultimately, it was the lowly third-class passengers who provided railway companies with one of their mainstay incomes. To begin with, railway managers did not see their task as providing travel facilities for the poorer population. Understandably perhaps, their visions were coloured by the existing clientele of the stage coaches and train fares were often set at the same per mile rates: the sheer speed of the railway and the consequent

savings the traveller registered in not having to purchase board and lodging on the road were seen as attractions enough. Thus, as the map above reveals, there was relatively little difference in the travel that a pound would buy when comparing outside stage coach and second-class rail. However, third-class rail travel extended the range considerably, and with Gladstone's Railway Act of 1844 which laid down specific minimum operating standards for third-class travel (including penny-a-mile fares), the pattern was set. By 1852 one-third of passenger traffic receipts on British railways derived from third-class travel. The popular consumption of rail travel became such that one-third of the nation's population (roughly six million people) were able to visit the Great Exhibition in London in 1851.

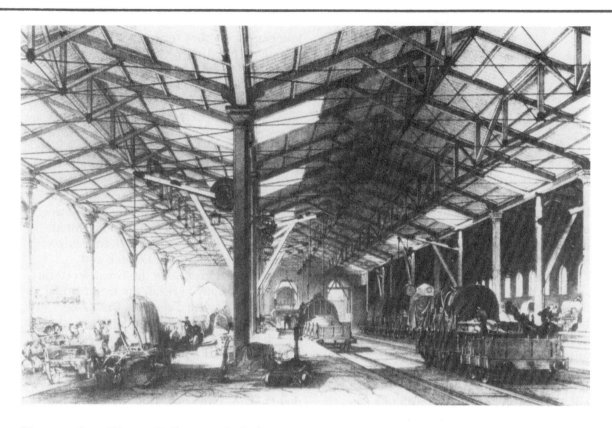

Plate 17: Great Western Railway goods shed, Bristol.

Compared to passenger traffic, the freight business developed slowly during the first decades of railway operation. Owing to the near parity in their charges, inland waterways were able to maintain a fairly strong hold on goods transport. They also had the organisation of the business under very effective control as a result of long practice, whereas the railways entered the field as relative novices. From 1847 the Railway Clearing House began to achieve a more effective business organisation for freight traffic, so that by 1850 receipts from freight handlings matched those from passengers, and a few years later railways eclipsed inland waterways as the principal movers of goods when measured in volumetric terms. There was no universal system of charges for railway freight traffic. Rates varied according to the type of traffic, whether goods were for export or for domestic consumption and according to the company area. Most companies with a distinct geographical identity sought to assist the trades of their regions and hence evolved a pricing policy accordingly.

CHARGES FOR GENERAL MERCHANDISE : 1842		
North Midland	2.24d	per ton per mile
Manchester & Leeds	3.1d	
Grand Junction	3.1d	
London and Birmingham	3.259d	
Great Western	3.685d	
Liverpool and Manchester	4.0d	
London and South Western	4.5d	

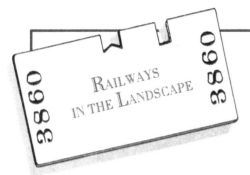

The railways brought to the landscape some highly novel elements. In non-material terms they spread the noises and smells of the machine age beyond the industrial valleys of midland and northern England. Contemporary artists might well record the unfamiliar clouds of steam wafting across pastoral landscapes, but they could not capture the clanking and hissing, the jarring of metal against metal and the oily, acrid smells so characteristic of the moving steam engine. In material terms the railways gave entirely new perspectives to the landscape. They cut great thrusting lines in a way only the Roman roads had done two millennia before. They gave birth to a new topography as cuttings, embankments, viaducts and bridges were created to maintain a level course. Canals anticipated the trend, but in their use of locks they bowed to existing topography. As they traversed the countryside, moreover, canals were anything other than symbols of the machine age — in the silent passage of horse-drawn barges and as water courses which merged into the surroundings much as the rivers from which they fed. Architects of early railway buildings sensed the powerful new symbolism of the working lines and evolved a classical-monumental style to echo it. The Moorish Arch at Edge Hill on the Liverpool and Manchester offers one of the earliest examples. The station at Lime Street was no less impressive. But the movement found its apotheosis in Philip Hardwick's portico at Euston for the London and Birmingham Railway. On the Great Western, Brunel was happy to recall the monumental tradition of English Tudor. At Bristol he created a train shed reminiscent of an Elizabethan banqueting hall. Even secondary lines shared in the phenomenon. For Gosport, Hampshire, William Tite built a massive colonnaded front. His chief justification was that Queen Victoria used the station on her way to Osborne House in the Isle of Wight.

Plate 18: Blisworth cutting, London and Birmingham Railway. Note the stone sleepers.

Plate 19: Hardwick's magnificent Doric portico at Euston.

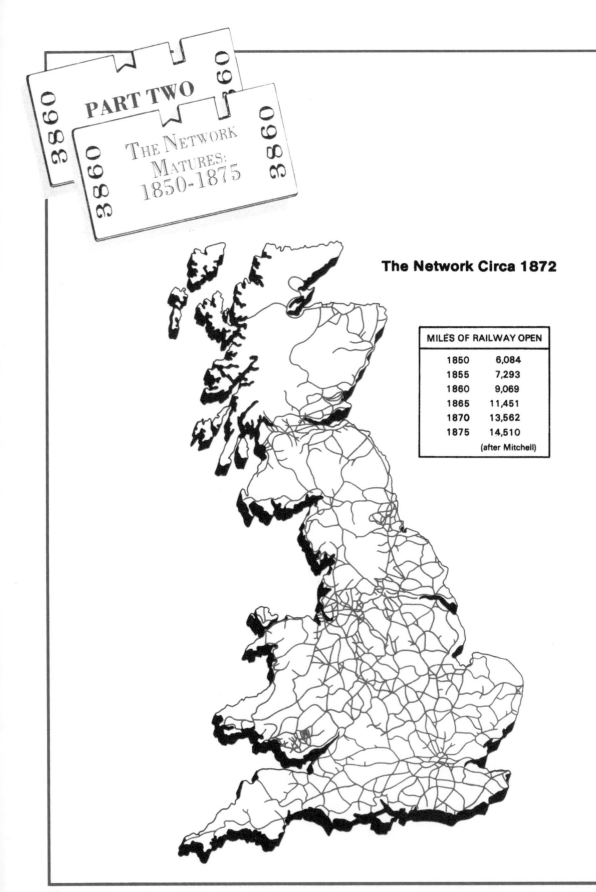

The Network Circa 1872

MILES OF RAILWAY OPEN	
1850	6,084
1855	7,293
1860	9,069
1865	11,451
1870	13,562
1875	14,510
	(after Mitchell)

Plate 21: Brunel's triumphal bridge across the River Tamar at Saltash — seen here nearing completion in 1859.

By the early 1850s the railway network had attained a degree of physical maturity such that few cities and towns of any importance were unable to boast a railway service. The areas remaining unserved were mostly confined to the peripheral parts of the country, notably Wales, northern Scotland and the West Country. In spite of this, much railway investment was yet to come. Roughly 60 per cent of the railway capital raised between 1825 and 1875 came after 1850. Between 1850 and 1875 some 8,500 miles were added to the network. The early 1860s witnessed an investment wave in some measure comparable to the earlier manias. Throughout the 1850s, gross railway capital formation had fallen back to between one and one and a half per cent of national income. By 1865 it had risen again to almost three per cent. The large-scale amalgamation of companies of the 1840s was matched in subsequent decades by further concentrations of management control, particularly in the relative hold that a select few companies exercised over gross revenue and capital. Each with capitalisations of over £20m, the LNWR, the GWR, the NER and the Midland were easily among the largest corporate business concerns the country had known. The birth of new independent companies did not cease by any means. But the share they held in the railway business was exceptionally small. In 1870, for instance, 83 per cent of total railway revenue accrued to only fifteen different companies. Such a high degree of concentration might initially be seen as reflecting a practical concern for cheap working and efficiency. In fact, it was much more a result of commercial power and empire building.

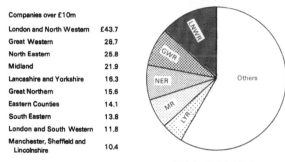

Railway Capital Authorised: 1858 (£m)

Companies over £10m	
London and North Western	£43.7
Great Western	28.7
North Eastern	25.8
Midland	21.9
Lancashire and Yorkshire	16.3
Great Northern	15.6
Eastern Counties	14.1
South Eastern	13.8
London and South Western	11.8
Manchester, Sheffield and Lincolnshire	10.4

Total Great Britain £324.5m

Plate 22: St Pancras station roof under construction.

Plate 23: The Midland Railway's London extension. Construction work at Mill Hill Lane Bridge, June 1867.

The network additions of this period involved intensification within areas already served as well as extensions into the remoter regions. The South Wales Railway linked Chepstow and Swansea by 1850 and extended to Haverfordwest by 1854. In north-east Scotland the great North of Scotland Railway opened its first lines in the mid-1850s. By 1859 London had a through rail connection to Penzance via the Cornwall Railway. Over the 1860s central Wales was slowly opened up to railways. In and around London, new railway lines were springing up almost annually and in 1863 the first underground line opened

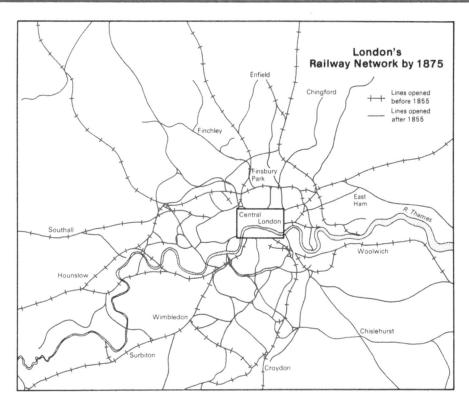

London's
Railway Network by 1875

Enfield

Chingford

Finchley

+—+—+ Lines opened
        before 1855
———— Lines opened
        after 1855

Finsbury
Park

East
Ham

R. Thames

Southall

Central
London

Woolwich

Hounslow

Wimbledon

Chislehurst

Surbiton

Croydon

between Bishop's Road near Paddington and Farringdon Street in the City. In 1868 the Midland Railway opened its London extension into St Pancras, marking the start of a long competitive struggle for traffic with its neighbour the London and North Western. The Victorians' passion for branch railway lines to almost every small town was still some years away, but by 1875 few parts of the country were far from rail communication. Indeed, the effect of main line extensions like that of the Midland was to provide the public with several alternative paths of railway travel between particular places.

Plate 24: The Great Western's London terminus at Paddington: the station hotel, another design by Hardwick, was erected between 1850 and 1852.

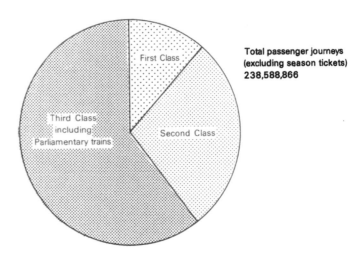

Composition of Passenger Traffic: 1865

First Class

Third Class including Parliamentary trains

Second Class

Total passenger journeys (excluding season tickets)
238,588,866

The dominating feature of the railway passenger business after 1850 was the growth of third-class travel. Gladstone's regulatory Act of 1844, which prescribed minimum facilities for low fare passengers, was eventually followed by an active policy on the part of the railway companies to pursue low-margin, high-volume business. By the mid-1860s almost two-thirds of passenger journeys were accounted for by third-class and Parliamentary (as of the 1844 Act) fares. Third-class passengers were still not catered for on all timetabled trains, however. Only after 1872, when the Midland Railway stole a march on its rivals and introduced third-class carriages on all its trains, did the poorer traveller begin to find access less restricted. The Midland set the pace again three years later by abolishing the second class altogether and raising the standards of accommodation in third-class carriages. The volume increase in railway passenger business is illustrated in the accompanying graph. In 1850 passenger journeys totalled some 60 million. By 1870 the total was almost 300 million. The rate of increase accelerated after the mid-1860s — partly as a result of the Midland Railway's competitive drive, but also because in 1871 the institution of holidays with pay (under the Bank Holidays Act) generated a vast new excursion business.

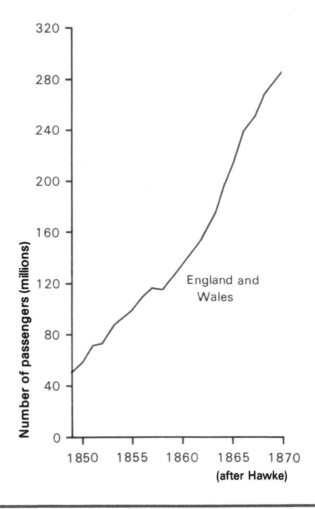

Passenger Journeys (millions)

Number of passengers (millions)

England and Wales

(after Hawke)

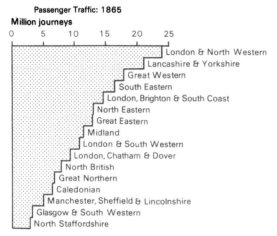

Passenger Traffic: 1865
Million journeys

London & North Western
Lancashire & Yorkshire
Great Western
South Eastern
London, Brighton & South Coast
North Eastern
Great Eastern
Midland
London & South Western
London, Chatham & Dover
North British
Great Northern
Caledonian
Manchester, Sheffield & Lincolnshire
Glasgow & South Western
North Staffordshire

Note : Companies below 2 million not shown

Plate 25: GWR broad gauge express passenger locomotive Tornado, *built in 1849.*

Plate 26: York station, with its grand curving overall roofs, was completed in 1877. In this view, taken around 1910, petrol-electric railcar no. 3170 waits at platform 5.

Plate 27: GWR broad gauge goods locomotive
Xerxes, of 1863.

Freight Tonnage (million tons)

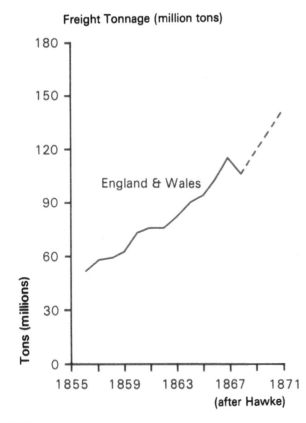

Whereas before 1850 freight took second place to passenger traffic on Britain's railways, after that date the freight business increasingly dominated and by 1865 yielded £19.3m of revenue against the £16.6m for the passenger business. Using a combination of pricing and takeover tactics, the railway companies were soon able to wrest control of inland freight from the waterways. At the same time, improved locomotive design and performance enabled them to tap entirely new traffics and to permit a vastly increased scale for existing ones. In 1850 the railways conveyed some 38 million tons of freight. By the mid-1870s the total was approaching 200 million. Pride of place went to the growth of mineral traffic, especially coal. In 1865 the North Eastern Railway carried almost four times as much mineral traffic as it did merchandise. Largely bereft of canals, the economy of north-eastern England was transformed by railways. Indeed, the North Eastern Railway, which came to serve the region almost exclusively, found in the carriage of coal its *raison d'être*.

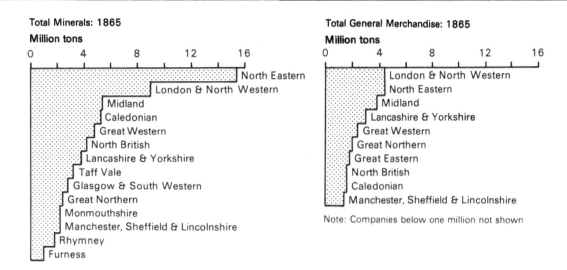

Total Minerals: 1865
Million tons

| 0 | 4 | 8 | 12 | 16 |

North Eastern
London & North Western
Midland
Caledonian
Great Western
North British
Lancashire & Yorkshire
Taff Vale
Glasgow & South Western
Great Northern
Monmouthshire
Manchester, Sheffield & Lincolnshire
Rhymney
Furness

Total General Merchandise: 1865
Million tons

| 0 | 4 | 8 | 12 | 16 |

London & North Western
North Eastern
Midland
Lancashire & Yorkshire
Great Western
Great Northern
Great Eastern
North British
Caledonian
Manchester, Sheffield & Lincolnshire

Note: Companies below one million not shown

Plate 28: Growth of traffic meant that constant improvements in station facilities were necessary and the LBSCR's experience at Lewes was typical. Opened in 1846, the station was resited in 1857 (seen here thirty years later) and reconstructed once more in 1889.

The Locomotive Stable

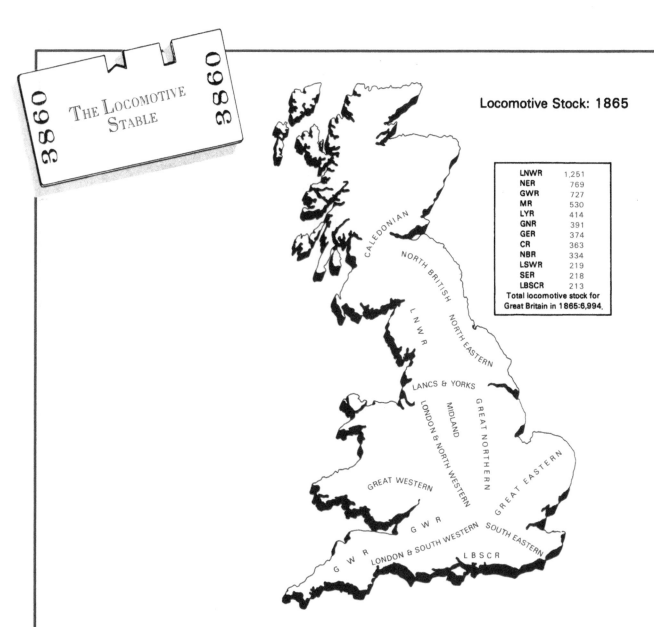

Locomotive Stock: 1865

LNWR	1,251
NER	769
GWR	727
MR	530
LYR	414
GNR	391
GER	374
CR	363
NBR	334
LSWR	219
SER	218
LBSCR	213

Total locomotive stock for
Great Britain in 1865:6,994.

When the first public railways were being inaugurated, steam traction remained at a highly experimental stage; the Rainhill locomotive trials which preceded the Liverpool and Manchester's opening were obvious enough testament. As a result, the early decades of railway operation witnessed a continuous process of design experiment. Locomotives varied in their basic size, in their overall centre of gravity, in the position of driving wheels in relation to boiler and firebox, in cylinder layout and, more widely, in general aesthetic appearance. After about 1850, however, certain standard design features were becoming clear. The most memorable was the preference for a single-driver wheel arrangement on fast passenger engines. Other passenger work was dominated by engines of a 2-4-0 pattern. For goods traffic the 0-6-0 wheel arrangement was most favoured. The London and North Western in fact built over 800 examples of one particular 0-6-0 design between 1858 and the early 1870s, a very early instance of standardisation on a British railway. As engines grew in size and tractive capacity, wheel arrangements naturally had to be extended. From the mid-1860s, the frontal bogie began to be featured in new locomotive designs, as did twin driving wheels in the increasingly favoured 4-4-0 arrangement.

Plate 29: LSWR Beattie-designed 2-4-0 no. 71 Alaric at Torrington. The locomotive entered service in 1863.

Plate 30: SER Cudworth 'mail' locomotive no. 72, built 1865.

The Railway Network
Circa 1900

By the mid-1870s, the main line network as we know it today was all but complete. Total route length was approaching 15,000 miles and the investment impact of railway construction was starting to diminish. In the 1860s boom, railways absorbed some 26 per cent of total domestic investment, but by the first decade of the twentieth century the proportion had fallen to under 10 per cent. Given the small geographical size of the country the scope for further main line additions to the network was obviously limited. Perhaps the most spectacular construction feature of the period was the extensive bridging and tunnelling in an effort to shorten routes. The Severn Tunnel, completed in 1886, greatly reduced journey times on the London to South Wales line. A year later a second Tay Bridge was opened (the first having collapsed during a storm in 1878), to be followed by the Forth Bridge in 1890, each of them contributing to faster schedules between Edinburgh, Dundee and Aberdeen.

If the really spectacular constructional feats belonged to the past, the railways were far from inactive during these last decades of the Victorian age. Indeed, route length expanded right through to 1912 to give a total of some 20,000 miles. Most of the new track consisted of branch and suburban lines, loop lines and cut-offs, many of which acted as feeders to the main line network or served outlying communities. Some were built by the larger companies as they vied with each other in size. Others were the creation of minor companies which sprang up to cater for the needs of particular localities, although in time these companies were absorbed by larger undertakings. The desire of even the smallest of urban settlements to possess a railway was an exceptionally powerful one. Few of these lines ever made a profit. But their absorption by contiguous companies helped to consolidate territorial monopolies and to identify the railways more closely with the communities they served.

Plate 31: The southern portal of the Severn Tunnel, seen soon after completion, dwarfs a contractor's locomotive.

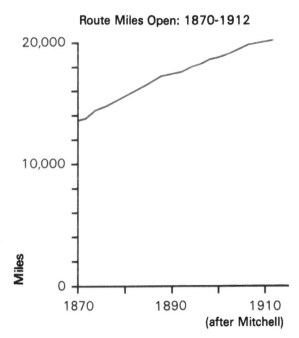

Plate 32: Manual and mechanical labour on the LSWR's 'New Line' to Guildford via Cobham in 1883.

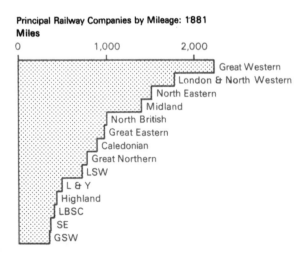

Principal Railway Companies by Mileage: 1881

Miles

0   1,000   2,000

Great Western
London & North Western
North Eastern
Midland
North British
Great Eastern
Caledonian
Great Northern
LSW
L & Y
Highland
LBSC
SE
GSW

The trend towards company amalgamation which had characterised the middle decades of the nineteenth century faded considerably in later years. The example of the North Eastern Railway, which in 1863 was able to boast no less than 37 former separate companies in its system, was not to become widespread until the turn of the century, and then most of the annexations were of small, single-route concerns. Attempts in 1871 to merge some of the biggest companies came to nothing, Parliamentary objections being the decisive factor. Later attempts at amalgamation often foundered because of entrenched boardroom enmities. Of the largest companies, the Great Western appeared to make most progress, particularly with its absorption of the Bristol and Exeter Railway in 1876 and the South Devon Railway in 1878. By 1881 it had the largest mileage of any British railway company.

Over the closing decades of the nineteenth century and up to 1914, the bulk of the railway system was thus operated by some dozen separate companies, not devoid of working agreements, but ever fearful of territorial infiltration or losses to their traffic. After years of bitter competition the South Eastern and the London, Chatham and Dover established a working union in 1899. But on the prestigious western route from London to Scotland competition was intensified when the Midland opened their famous Settle and Carlisle line in 1876, completing a new through route to the Borders. Some twenty years later, the Midland itself became a competitive target when the Manchester, Sheffield and Lincolnshire (retitled the Great Central from 1897) built the last main line into London providing further rivalry for the lucrative traffics of the English heartlands.

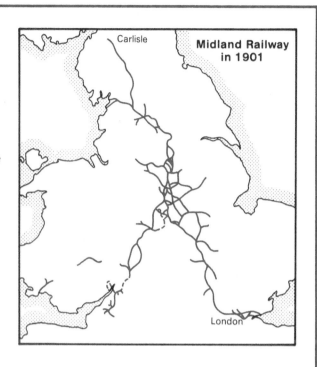

Plate 33: The Midland Railway's main line through the Peak District: a construction scene near Grindleford.

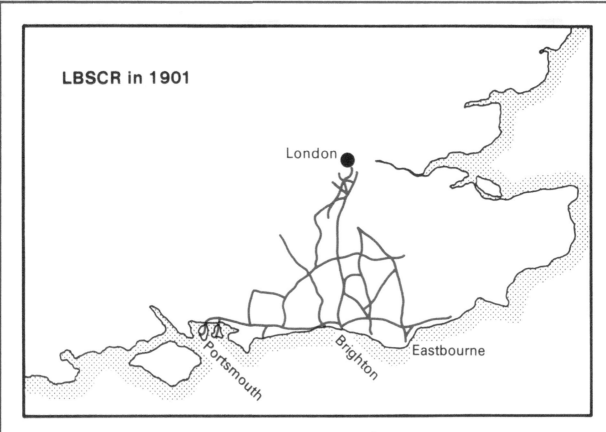

LBSCR in 1901

The British railway map was not neatly divided among the major railway companies. A few concerns, for instance the London, Brighton and South Coast and the North Eastern, operated in distinct geographical areas, but others, such as the Midland, had lines extending almost the length of the country. Certainly the Midland had an important network of lines in the areas after which it was named, but by 1901 its influence penetrated South Wales and Somerset in the west, Essex in the east, and north to the borders of Scotland. If the company's Irish steamship routes from Heysham harbour are included, the degree of penetration was even greater. The lines of the London and North Western described a very similar pattern, so helping to explain why the two companies became such arch-rivals. Some railways fitted neither the pattern of the LBSCR nor that of the Midland. One example was the Great Central Railway. Starting life as the Manchester, Sheffield and Lincolnshire Railway with a somewhat paltry and disconnected set of lines between the

Humber and the Mersey, its managers later embarked on a grandiose yet singular London extension. The resulting network resembled a T-shape, the London line appearing like some umbilical cord to economic salvation; it never was.

Great Central in 1901

Plate 34: Horsted Keynes station in 1882 (LBSCR).

Plate 35: The concourse at Marylebone, London terminus of the Great Central Railway, circa 1899.

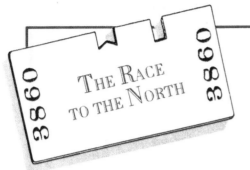

The Racing Routes

—— East Coast Route
(GNR, NER, NBR)
—— West Coast Route
(LNWR, CR)

Aberdeen

Edinburgh

Newcastle
Carlisle

York

Crewe

LONDON

The number of independent railway companies in Britain and the competition that arose among them sometimes conferred enormous benefits on the travelling public, particularly as far as the speed of travel was concerned. The effect was especially marked on the major long distance routes, none more impressive than the east and west coast main lines. In 1852 Anglo-Scottish expresses took eleven to twelve hours on the journey from London to Edinburgh or Glasgow, and this remained much the pattern for the next twenty years. The principal companies involved were the Great Northern and the London and North Western, responsible as they were for the southern sections of the routes. The catalyst for an improvement in schedules was the Midland's entry into London in 1868 and its de-regulation of passenger class divisions, leading ultimately to abolition of the second class. The Great Northern and its east coast partners responded first, but by the mid-1880s east and west coast companies were engaged in a speed war. The climax came in August 1888 when the two sides were openly racing for the title of fastest to Scotland. The journey time fell to almost 7½ hours. On 14th August, 1888, the companies agreed to a truce and the east coast schedule was set at 7¾ hours, the west coast at 8 hours. Within seven years, though, the race was renewed, except that Aberdeen instead of Edinburgh became the ultimate goal. Over six weeks in July/August 1895 the journey time from London to Aberdeen plummeted from 11½ hours to 8½. As one railway commentator put it, by mid-August Aberdonians were rubbing their eyes and wondering if the granite city was soon to become a kind of northern suburb of London.

This map provides a graphic demonstration of the radically improved schedules achieved on the Anglo-Scottish rail routes over the last quarter of the nineteenth century. The railways effectively shrank linear distance by about a third.

Shrinking National Space

E

Edinburgh

LONDON

THE

BEST TRAINS.

PARTICULARS

OF THE

Summer Services of 1888,

AND

THE RAILWAY RACE to EDINBURGH.

"The spirit of the time shall teach me speed."—
SHAKESPEARE: *King John*, Act iv., Scene 2.

"PALL MALL GAZETTE" OFFICE, 2, NORTHUMBERLAND STREET, STRAND, LONDON, W.C.

Plate 36: Crewe panorama, 1895; steel foundry in centre.

Perhaps commensurate with their enormous scale as business concerns, the railways in Britain spawned their own urban colonies. The construction and maintenance of rolling stock for operation over many miles of network required both a vast army of men and an enormous area of industrial workshops. Over the course of the nineteenth century, therefore, many of the larger companies founded their own towns — usually on green-field sites and with full provision for workers' housing. Crewe was the most famous of these company towns, built by the Grand Junction Railway (later the LNWR) in an unpopulated area of Cheshire in the 1840s. Crewe was a pioneer in urban welfare, the company providing recreational and health facilities, savings schemes, as well as piped water and sewage disposal. The example of Crewe was imitated at New Swindon (GWR) and at Wolverton, Bucks (London and Birmingham Railway). By the close of the century Crewe had grown into a town of over 40,000 people with an economic momentum of its own.

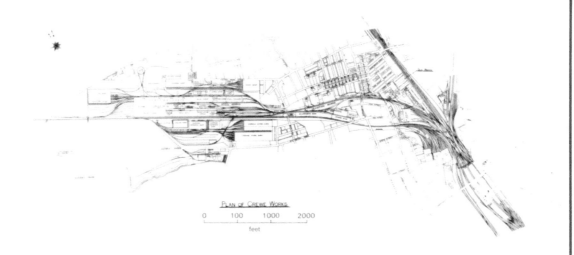

PLAN OF CREWE WORKS

0   100   1000   2000

feet

Plate 37: The Fitting Shop, Crewe Works.

*Plate 38: The fantastic neo-Gothic facade of
St Pancras. The Midland Railway's London
terminus is seen here in the 1920s.*

New towns such as Swindon and Crewe were
only one aspect of the railways' impact upon
the Victorian urban scene. In many existing
towns they transformed the nature and scale of
economic and social development through the
ability to assemble and despatch materials and
products in a way never possible with previous
means of transport. For many commentators,
however, it was the impact of the railway upon
the structure and life of the great Victorian
cities which provided its most potent
expression. Whereas before 1850 companies
were often content to make their termini on
the fringes of the existing built-up areas, by
the closing decades of the century they had
infiltrated to the very cores, clearing enormous
tracts of urban land in their path so as to
make way for passenger and goods terminals,
locomotive depots and marshalling yards, and
the veritable tangle of junctions and connecting
lines essential to efficient operation. By 1890
roughly £100m had been expended on railway
terminals. In many cities, railways owned up to
eight or ten per cent of the central land. More
broadly they had become unconscious arbiters
of urban plans, with their wide approach tracks
acting on the one hand as near-impenetrable
barriers to movement, and on the other as
bases for the layout of streets and sometimes
whole quarters. In some cities, terminals were
closely juxtaposed, as with King's Cross and
St Pancras in London. And companies like the
Midland added to their areal domination of the
central districts by building terminal façades of
considerable vertical scale and architectural
power.

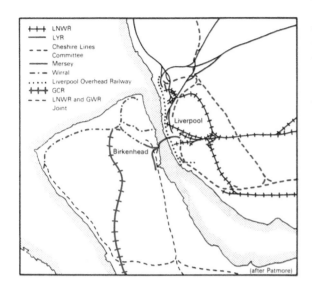

Legend:
- +++ LNWR
- —— LYR
- - - - Cheshire Lines Committee
- —— Mersey
- -·-·- Wirral
- ······ Liverpool Overhead Railway
- +++ GCR
- - - - LNWR and GWR Joint

Liverpool

Birkenhead

(after Patmore)

Liverpool's railway experience was typical of the large conurbations by the early twentieth century. Both banks of the Mersey were criss-crossed with lines of different companies. Companies that failed to gain access to Liverpool itself found in Birkenhead an easy alternative entry.

Plate 39: The railway sweeps all before it:
St Pancras station ground in August 1867.

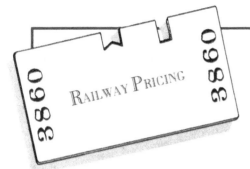

RAILWAY PRICING

Railway Rates on Timber and Deals: 1865

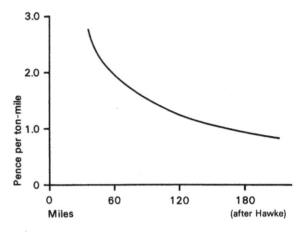

Pence per ton-mile

3.0

2.0

1.0

0

0   60   120   180

Miles   (after Hawke)

Though the railways cheapened the price of transport appreciably, traders never ceased to complain about railway charges. This dissatisfaction grew as the railway companies grew larger and more powerful and competition in price diminished. Ultimately it led to far-reaching legislation to control freight charges, introduced between 1888 and 1894. One of the major weaknesses of British railway policy in the nineteenth century was the failure to develop a scientific and rational pricing policy. Each Act authorising the establishment of a company simply prescribed maximum charges within which the railway concerned was free to fix its charges on the basis of a classification drawn up by the company and later consolidated by the Railway Clearing

House. In time, with the proliferation of companies, the situation became extremely complex. By the 1880s a multiplicity of charges were levied under no less than 900 Acts of Parliament. Under the new legislation over the period 1888 to 1894, uniform scales of rates and a standard classification of commodities were substituted. Maximum charges were fixed and companies had to justify any applications for an increase on the basis of a change in the costs of operation.

Plate 40: The Midland Railway's Avon Wharf at Bristol in June 1898.

Plate 41: Great Northern Railway open third, circa 1905.

On British railways, both passenger and freight charges were fixed largely on what the traffic would bear rather than on the cost of service rendered, and some companies even set commodity rates so as to favour the trade of the regions they served. For passenger traffic, the policy concentrated on attracting volume business, with an increasing pursuit of the lower end of the passenger market as the nineteenth century progressed. This chiefly relied on exploiting the price-elasticity of potential rail travellers, though some care was taken to uphold standards of comfort and service. The ability of this strategy to yield satisfactory economic returns was a secondary consideration.

In the mid-1860s third-class passengers paid on average just over one penny a mile, but choice of trains was restricted and comfort left much to be desired. By the early 1900s, cheap and concessionary fare policies had brought the figure down by half, while standards of accommodation and running had become markedly superior. All the time railways held a near monopoly of internal transport, such pricing policies proved workable and financially viable. But when competition from motor transport began, the policies became a liability.

Reported Rates per Passenger Mile: Autumn 1866		
First Class	Express	2.80d
	Ordinary	2.25
Second Class	Express	2.03
	Ordinary	1.68
Third Class		1.02

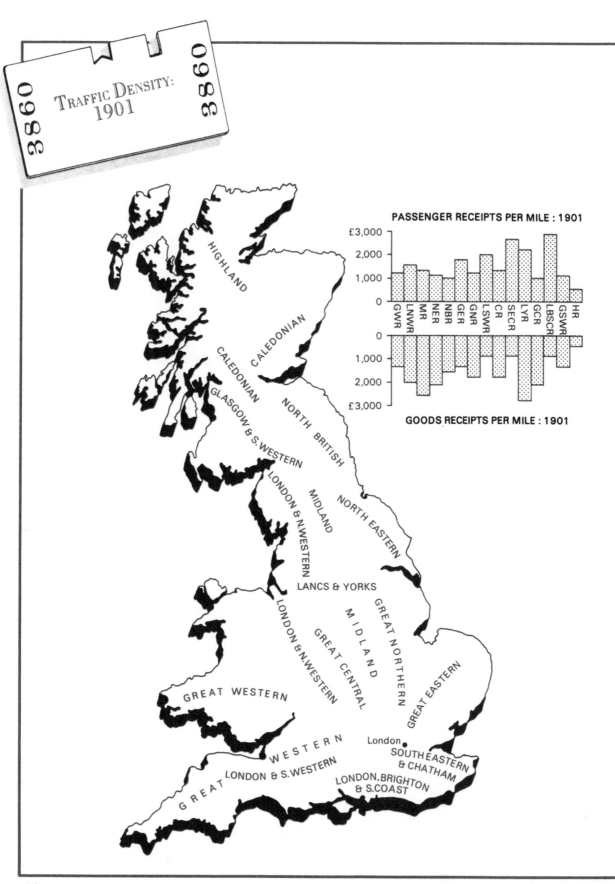

PASSENGER RECEIPTS PER MILE : 1901

£3,000
2,000
1,000
0

GWR LNWR MR NER NBR GER GNR LSWR CR SECR LYR GCR LBSCR GSWR HR

0
1,000
2,000
£3,000

GOODS RECEIPTS PER MILE : 1901

HIGHLAND

CALEDONIAN

CALEDONIAN

GLASGOW & S.WESTERN

NORTH BRITISH

LONDON & N.WESTERN

MIDLAND

NORTH EASTERN

LANCS & YORKS

MIDLAND

GREAT NORTHERN

GREAT CENTRAL

GREAT EASTERN

LONDON & N.WESTERN

GREAT WESTERN

WESTERN

London

SOUTH EASTERN
& CHATHAM

GREAT

LONDON & S.WESTERN

LONDON, BRIGHTON
& S.COAST

Plate 42: LSWR express, hauled by Adams T6 4-4-0 no. 685.

The size of territory served by a railway company was seldom a guide to the density of its traffic. The Great Western company, although having one of the largest spheres of influence, ranked comparatively low in traffic receipts per track mile. The Lancashire and Yorkshire, by contrast, presented an entirely reverse picture: a limited territorial coverage but high goods and passenger receipts. The difference reflected the relative densities of route patterns as well as the volume of traffic over them. The Great Western system was widely diffused and had much single line which carried a restricted traffic. The Lancashire and Yorkshire had a greater proportion of double- and multiple-tracked route and offered a tightly knit lattice of rail communication.

Average receipts per track mile in the United Kingdom in 1901 were £2,764, of which goods constituted £1,470 and passengers £1,294. The companies with the heaviest passenger traffic were the London, Brighton and South Coast, and the South Eastern and Chatham. Then, as now, these lines carried a large commuter traffic. In season they also carried hoards of holiday makers to the resorts of the Kent and Sussex coasts. The lowest traffic was recorded on the Highland, where population was sparse and where lines were never really built to yield investment profits. As regards goods, the Lancashire and Yorkshire and the Midland held the prime positions, both serving thriving mining and manufacturing districts; and, with the London and North Western and Great Central companies, holding a virtual monopoly of the industrial heartlands. Predictably, goods receipts were least in the agricultural south, though none as low as the Highland Railway where, to the disarming critic, trains might be said to have been more common than traffics.

Plate 43: SECR local passenger train with 2-4-0 no. 262 at its head.

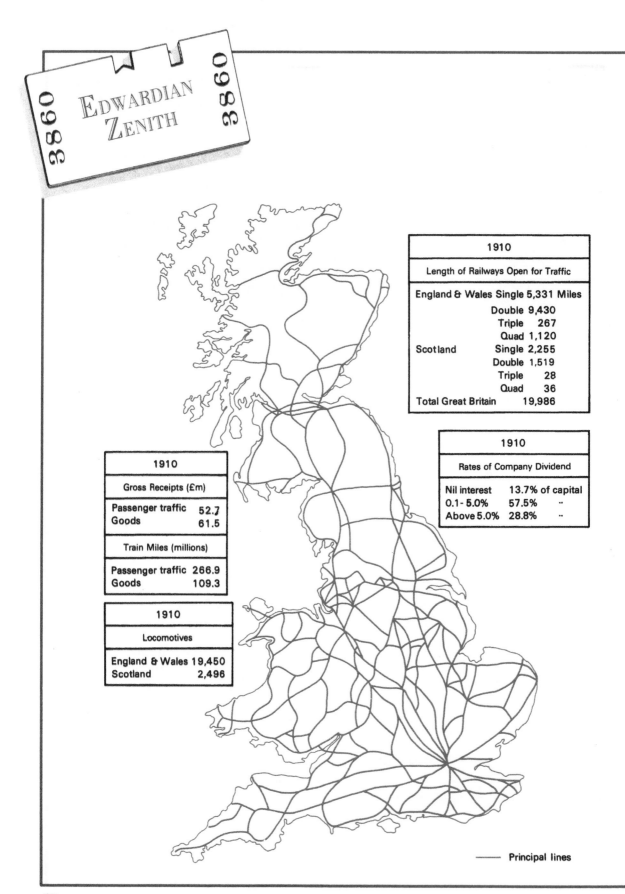

1910
Length of Railways Open for Traffic

England & Wales	Single	5,331 Miles
	Double	9,430
	Triple	267
	Quad	1,120
Scotland	Single	2,255
	Double	1,519
	Triple	28
	Quad	36
Total Great Britain		19,986

1910
Rates of Company Dividend

Nil interest	13.7% of capital
0.1- 5.0%	57.5% "
Above 5.0%	28.8% "

1910
Gross Receipts (£m)

Passenger traffic	52.7
Goods	61.5

Train Miles (millions)

Passenger traffic	266.9
Goods	109.3

1910
Locomotives

England & Wales	19,450
Scotland	2,496

—— Principal lines

*Plate 44: Doncaster GNR — north-facing view of
the carriage-building shops on Ledger Day,
13th September, 1911.*

The reign of Edward VII saw the railways of
Britain at their zenith. By the close of the
Victorian age they dominated the transport
scene to an unprecedented degree, leaving
waterways, coastal shipping and road transport
to act as marginal suppliers. In 1905 freight
traffic carried by the Great Western Railway
alone was greater than the tonnage carried by
all the inland waterways of the country.
Between 1870 and 1912, the volume of freight
traffic increased approximately three-fold
and passenger carryings rose four-fold. The
railways did much to meet up the pent-up
demand for more and better passenger facilities
occasioned by the expansion in industrial
activity, rising real incomes, the spread of
suburban development and the desire for
greater mobility. The steady increase in the
significance of third-class travel which
characterised the middle decades of the
century continued unabated in later ones. By
the early 1900s all but five per cent of
passengers were travelling third class and the
average fare had dropped to ½d per mile.
Companies such as the North Eastern Railway
set the pace by introducing a wide range of
cheap travel facilities, including weekend and
tourist tickets, excursion facilities, and
reduced fares for workmen and servicemen.
Railway travel became the province of the
common man.

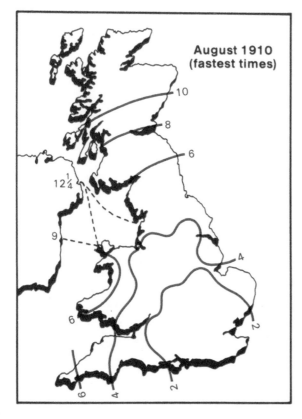

August 1910
(fastest times)

This map makes for an interesting comparison with the map for 1845. It was largely on the north-south routes that journey times had been improved in any real measure.

The closing decades of Queen Victoria's reign also saw a remarkable improvement in the frequency of railway services, in their speed and in standards of comfort and safety. Not all lines and not all companies followed the pattern, of course, but on the major routes of the larger companies rail travel was undeniably at its zenith. The 1870s saw the introduction of the first Pullman carriages, sleeping and dining cars followed, and the use of corridor coaches transformed the crack expresses into hotels on wheels. By 1910 the provision of meals on long-distance trains was almost universal, available to third-class as well as first-class passengers. A glance at Bradshaw's

Plate 45: Caledonian Railway express at Glasgow Central. The carriages are some of the sumptuously appointed West Coast Joint Stock for Anglo-Scottish services.

Plate 46: Great Northern Railway first class dining car, 1912.

famous railway guide illustrates the pattern with many of its train columns headed 'Breakfast Car', 'Luncheon Car' or 'Dining Car' Express. The more important trains were provided with steam heating, gas or electric lighting, washing facilities and lavatories, although the latter varied alarmingly.

Traders, too, benefited from better services and lower rates — though in the light of their constant complaints perhaps their requirements were not met so assiduously as those of passengers. Certainly the railways' methods of handling traffic have been much criticised by later commentators.

Plate 47: GWR TSS St Patrick *on the Fishguard-Rosslare service. A triple-screw turbine vessel, she was introduced in 1906.*

Plate 48: SECR boat train, 1908, hauled by a Wainwright 4-4-0.

The improvement of travelling facilities up to 1910 naturally absorbed considerable amounts of capital, particularly for new rolling stock and new locomotives. However, it should not be overlooked that the railways made similarly large investments in ancillary facilities, which included the construction and acquisition of docks, harbours and piers, steamships, road vehicles, canals, hotels, electric power stations and even an amusement park in Cleethorpes. The railways became the chief dock owners in the country, as well as operating one of the largest fleets of short-sea steamboats. Many of the latter were employed on the cross-channel ferry services which had reached a high standard by the end of the century.

Plate 49: Grangemouth Docks, Caledonian Railway.

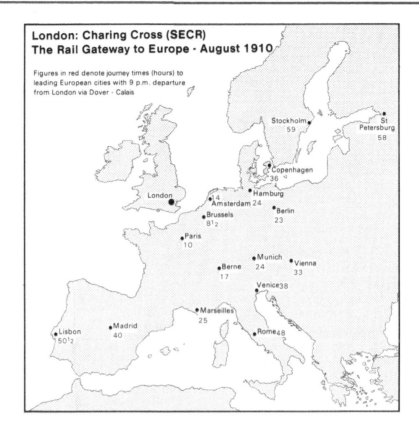

London: Charing Cross (SECR)
The Rail Gateway to Europe - August 1910

Figures in red denote journey times (hours) to
leading European cities with 9 p.m. departure
from London via Dover - Calais

Stockholm 59
St Petersburg 58
Copenhagen 36
Hamburg 14
Amsterdam 24
London
Brussels 8½
Berlin 23
Paris 10
Munich 24
Berne 17
Vienna 33
Venice 38
Marseilles 25
Lisbon 50½
Madrid 40
Rome 48

Plate 50: Folkestone, SECR, with triple-screw steamer The Queen *leaving for Boulogne.*

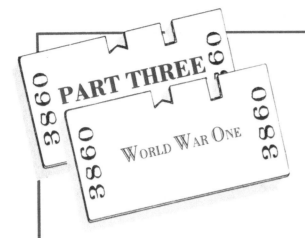

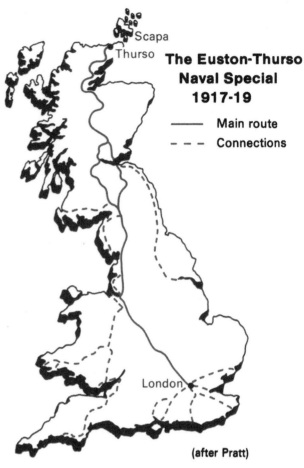

Scapa

Thurso

**The Euston-Thurso
Naval Special
1917-19**

——— Main route

– – – Connections

London

(after Pratt)

*During the war various sections of the British
navy were based on the eastern and northern
shores of Scotland and at Scapa in the
Orkneys. To cope with the resulting movements
of officers and men, a special train was
eventually established to run every weekday.
Between February 1917 and April 1919, some
470,000 persons were conveyed on it. The
train generally consisted of fourteen vehicles,
including sleeping cars.*

In 1914 the railway companies were obliged to
put their organisations at the disposal of the
government for the movement of troops and
supplies, in return for which they were
guaranteed a net income equal to that of 1913.
The railway system was operated on a unified
basis under the direction of the Railway
Executive Committee, a body composed of the
general managers of the main companies. The
railways had to cope with a much greater
volume of traffic during hostilities at a time
when their resources were depleted. Apart
from ordinary civilian traffic, they had to carry
troops, munition workers and supplies for the
war effort, while much of the traffic that before
the war had gone by water was diverted to the
railways for reasons of security and economy.
The Great Northern, for example, became a
major conveyor of war traffic: its goods traffic
increased by 125 per cent and 60 per cent of
its passengers consisted of servicemen. The

largest movement of troops was probably made
by the London and South Western since it
provided the main link to the Continent
through Southampton. That company carried
no less than 20 million soldiers throughout the
war. The Great Eastern was also very busy in
this respect, carrying 10½ million troops for
whom it provided 13,000 special trains, while
a further 11,000 'specials' were laid on to
handle other government traffic.

Despite the immense burden of the war few
railways broke down under the strain and the
task was accomplished with less manpower and
rolling stock and with a minimum loss of life
and injury. Yet about 30 per cent of the
pre-war staff had been released for national
service and considerable quantities of rolling
stock had been sent to the war theatres. That

the railways were able to perform such heroic feats under these conditions was due in no small part to the benefits flowing from the unified system of operation. Great efforts were made to rationalise services and reduce non-essential movements. To discourage civilian traffic, fares were raised, cheap travel facilities were withdrawn and appeals were made to the public to avoid unnecessary rail journeys. Competitive duplication of services between the same towns was curtailed by comprehensive reorganisation. As a result the number of passenger trains was reduced by nearly 40 per cent with considerable savings in manpower and rolling stock. The efficiency of freight handling was also improved by eliminating the light loading of wagons and unnecessary haulage, pooling the wagon stock and regulating the flow of goods at terminals. The benefits of wartime control prompted the

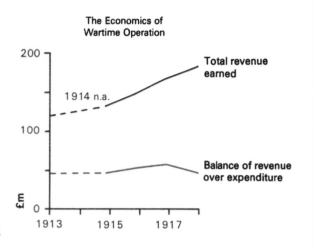

The Economics of Wartime Operation

Total revenue earned

Balance of revenue over expenditure

government to consider reorganising the railway system after the war. This did not fully compensate the railways for the high cost of their wartime efforts. By the end of the war the railways were in poor physical shape and their financial position was steadily deteriorating at a time when they were to face major new challenges.

Plate 51: A train of repatriated prisoners-of-war ready to depart from the SECR's Dover Pier station.

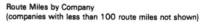

Route Miles by Company
(companies with less than 100 route miles not shown)

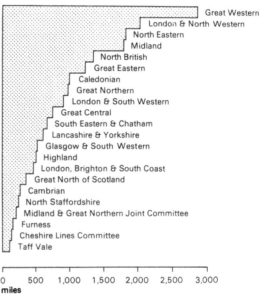

Great Western
London & North Western
North Eastern
Midland
North British
Great Eastern
Caledonian
Great Northern
London & South Western
Great Central
South Eastern & Chatham
Lancashire & Yorkshire
Glasgow & South Western
Highland
London, Brighton & South Coast
Great North of Scotland
Cambrian
North Staffordshire
Midland & Great Northern Joint Committee
Furness
Cheshire Lines Committee
Taff Vale

0   500   1,000   1,500   2,000   2,500   3,000
miles

Although the railways remained under government control from the Armistice until 1921, the different companies regained their distinctive identities as the various exigencies of war subsided. Fragmentation of ownership was almost as strong in 1921 as it had been before 1914. The astonishing array of liveries visible before the war at joint stations like Carlisle Citadel may have diminished, but the variations in working practice and in locomotive and stock designs were unchanged. Britain still did not have a unified railway system but an aggregation of systems, some with a strong territorial identity, some with networks that sprawled and intertwined across their neighbours' domains. One result was that customers faced a very wide choice of facilities, particularly on the main line routes. Unfortunately, however, what was of benefit to the public was often uneconomic to the company, and unwarranted duplication of routes and facilities was to extract a heavy price from company profits.

Plate 52: MR Somers Town goods depot in 1922, with St Pancras in the background.

Passenger Traffic in 1921

The contribution to passenger conveyance made by the different railway companies in 1921 varied enormously. This variation was partly a function of differences in route mileage. But it was also related to the density of population in the territory served. Predictably, perhaps, the Great Western had the highest mileage in 1921 and recorded the highest number of passenger journeys. The Metropolitan, by contrast, had a miniscule mileage but an intensive traffic, located as it was in London commuterland. Many of the Scottish companies presented a further contrast. The North British, the Caledonian, the Glasgow and South Western, and the Highland all had passenger traffics which were disproportionately low in relation to their mileages. In spite of the dense traffic in and around Glasgow, most Scottish lines catered simply for long-distance travel through sparsely populated uplands.

A surprisingly high number of passenger journeys were made on workmen's tickets — special concessionary fares which Parliament had compelled the railways to introduce in the latter part of the nineteenth century. Companies serving major industrial areas stand out on the graph, but so do those serving London with its armies of clerks and tradespeople.

Passenger Journeys (ordinary fares)

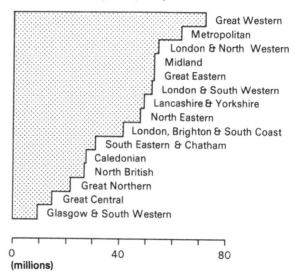

Passenger Journeys (workmen's fares)

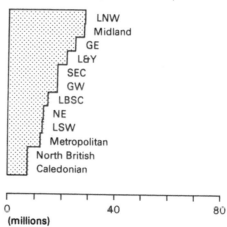

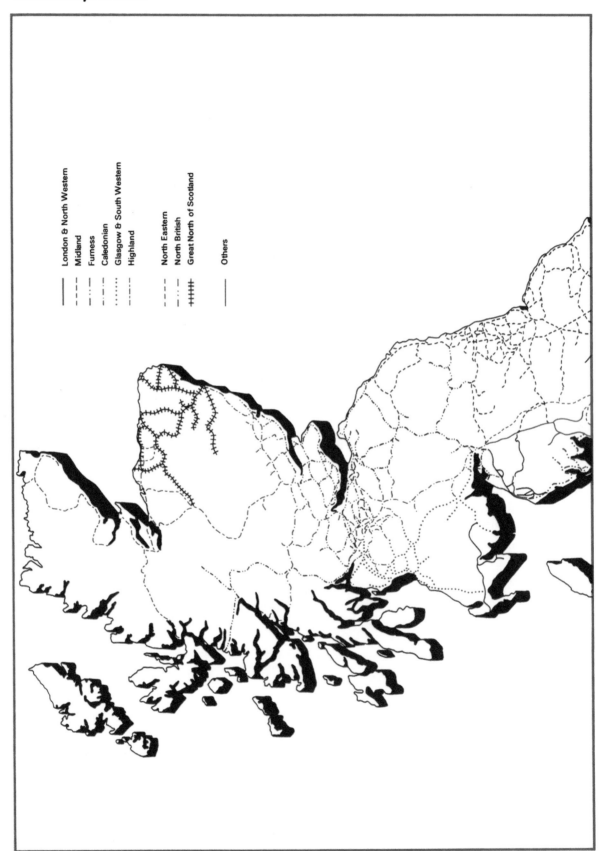

London & North Western

Midland

Furness

Caledonian

Glasgow & South Western

Highland

North Eastern

North British

Great North of Scotland

Others

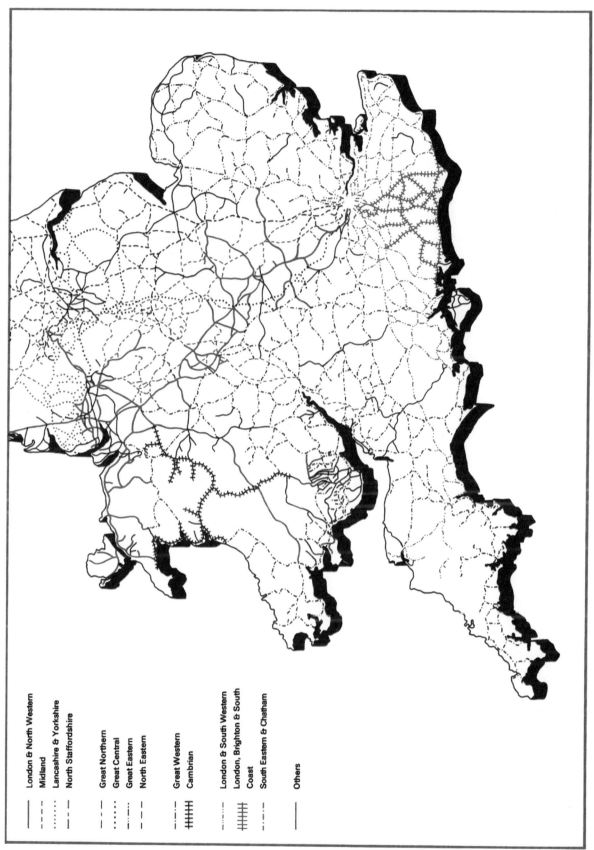

London & North Western
Midland
Lancashire & Yorkshire
North Staffordshire

Great Northern
Great Central
Great Eastern
North Eastern

Great Western
Cambrian

London & South Western
London, Brighton & South
Coast
South Eastern & Chatham

Others

Plate 53: A colliery in Scotland circa 1920. The wagons in view were colliery-owned but could be moved anywhere within the country's railway system.

Plate 54: A coal train on the LBSCR, hauled by E3 0-6-2T no. 461.

The Coal Trade in 1921

In the height of the canal era, the canal commentator John Phillips wrote that every successful inland navigation had coal at its heels. Had he been alive a century later, he could have said the same of railways with almost equal justification. For over much of their life coal provided the railways with their most lucrative traffic. Companies like the Taff Vale did virtually nothing else but convey coal. Larger railways like the North Eastern may have been less exclusive carriers but they made up for it in the vast scale of their coal traffic.

Little did they know it, but this whole foundation of railway prosperity was soon to collapse. British coal exports were to halve between 1913 and 1938 in the face of trade depression and the growth in oil-fired ships. Domestic demand was to slump, too, as coal-using industries like iron and steel contracted. However, such grim prospects were far from the minds of those who found themselves deliberating on the railways' future in 1921.

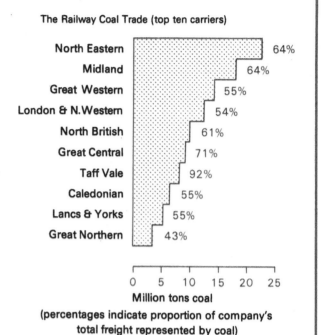

The Railway Coal Trade (top ten carriers)

North Eastern 64%
Midland 64%
Great Western 55%
London & N.Western 54%
North British 61%
Great Central 71%
Taff Vale 92%
Caledonian 55%
Lancs & Yorks 55%
Great Northern 43%

Million tons coal
(percentages indicate proportion of company's total freight represented by coal)

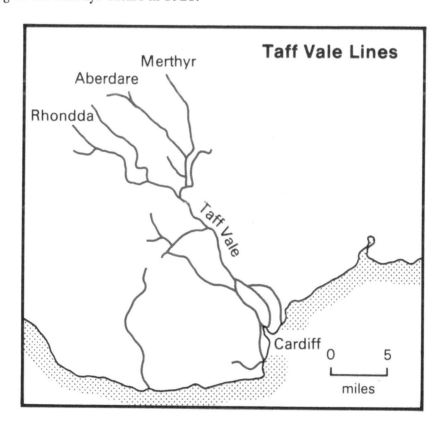

Taff Vale Lines

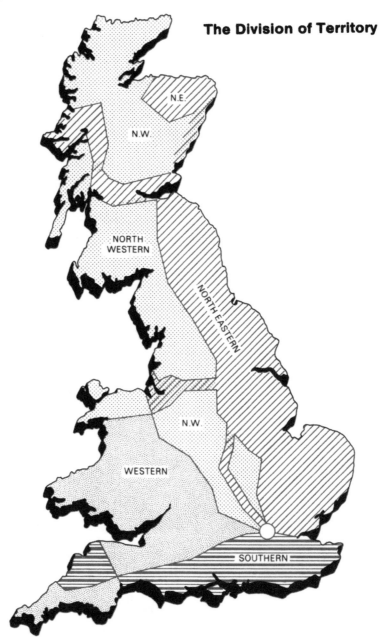

The Division of Territory

N.E.

N.W.

NORTH WESTERN

NORTH EASTERN

N.W.

WESTERN

SOUTHERN

Various schemes of territorial reorganisation were suggested when the grouping of the railway companies was under Parliamentary scrutiny. This was the one finally chosen. It dispensed with an earlier idea to operate the railways of Scotland under a separate company.

CONSTITUENT COMPANIES	route miles
LMS 7,790 route miles*	
London & North Western ⎫ Lancashire & Yorkshire ⎬	2667 $\frac{1}{2}$
Midland	2107 $\frac{3}{4}$
North Stafford	220 $\frac{3}{4}$
Furness	158
Caledonian	1114 $\frac{1}{2}$
Glasgow & South Western	493 $\frac{1}{2}$
Highland	506
LNER 6,590 route miles*	
North Eastern ⎫ Hull & Barnsley ⎬	1864 $\frac{1}{4}$
Great Northern	1051 $\frac{1}{4}$
Great Eastern	1191 $\frac{1}{4}$
Great Central	852 $\frac{1}{2}$
North British	1378
Great North of Scotland	334 $\frac{1}{2}$
GWR 3,800 route miles*	
Great Western	3005
Barry	68
Cambrian	295 $\frac{1}{4}$
Cardiff	11 $\frac{3}{4}$
Rhymney	51
Taff Vale	124 $\frac{1}{2}$
Newport & South Wales	10 $\frac{1}{4}$
SOUTHERN 2,200 route miles*	
London & South Western	1020 $\frac{1}{2}$
London, Brighton & South Coast	457 $\frac{1}{4}$
South Eastern & Chatham	637 $\frac{3}{4}$

* includes leased and worked lines and share of joint lines

When the railways were released from government control in August 1921 the companies were not allowed to revert to their pre-war competitive position. There was fairly general agreement that the benefits of unified control should be retained in peacetime and with this end in view the Railways Act was passed in that year. This was one of the most important pieces of railway legislation in the history of the industry since it represented the first attempt to grapple with the problems of railway organisation and planning as a whole. It provided for the amalgamation of some 120 separate companies into four large groups: the London and North Eastern (LNER), the London, Midland and Scottish (LMS), the Southern (SR) and the Great Western (GWR). These four groups remained in being until the railways were nationalised after World War Two.

Plate 55: The grouping arrangement did little to reduce the duplication of station facilities inherited from the former independent companies. This is Leeds Wellington station, Midland Railway, in 1903.

The grouping was achieved with great rapidity, and by 1923 the reorganisation had been completed. In effect it brought to a logical conclusion the trend towards company concentration evident in the late nineteenth and early twentieth centuries. The chief idea behind the amalgamation was that it would eliminate inter-company rivalry and reduce the cost and inefficiency resulting from the duplication of facilities. It was also envisaged that substantial economies would accrue from more unified operation. A second part of the 1921 Act dealt with charges and profits. The chief aim was to codify and simplify the complex system of charging and to provide a more flexible rate-making structure.

A permanent charging body, the Railway Rates Tribunal, was set up whose first duty was to approve a new schedule of standard charges submitted by the amalgamated companies. Although the Railways Act did not guarantee the companies a certain level of profit, it did require that charges in the first instance should be fixed at a level to yield an annual net revenue equivalent to that of 1913.

The Big Four

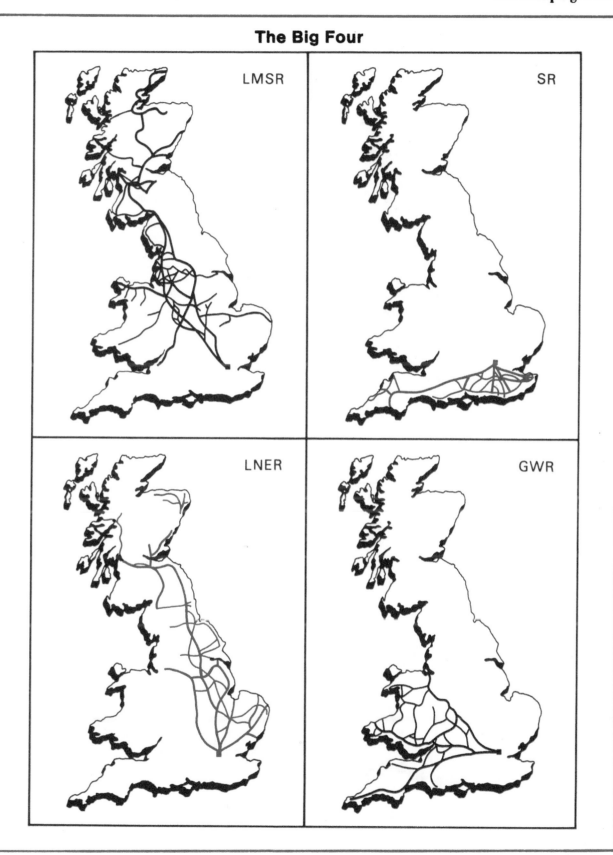

The Grouping: 1923

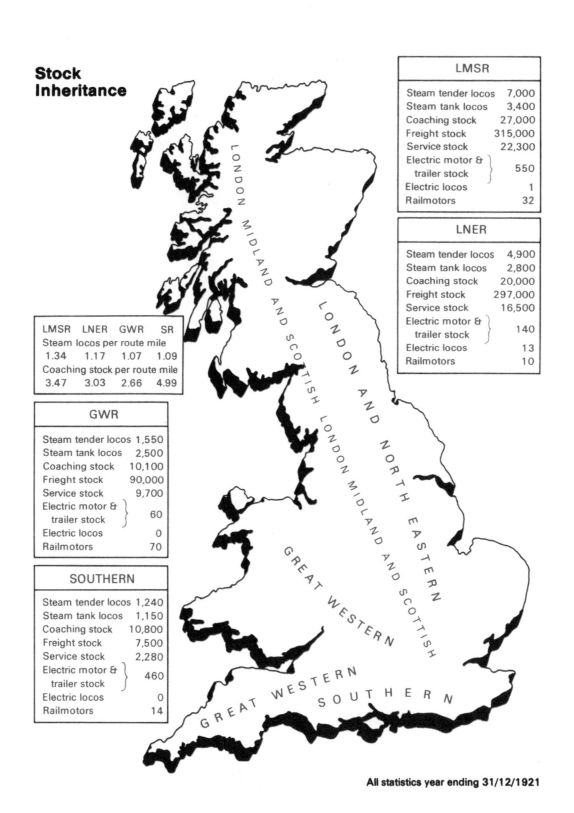

Stock Inheritance

	LMSR	
Steam tender locos		7,000
Steam tank locos		3,400
Coaching stock		27,000
Freight stock		315,000
Service stock		22,300
Electric motor & trailer stock		550
Electric locos		1
Railmotors		32

	LNER	
Steam tender locos		4,900
Steam tank locos		2,800
Coaching stock		20,000
Freight stock		297,000
Service stock		16,500
Electric motor & trailer stock		140
Electric locos		13
Railmotors		10

	LMSR	LNER	GWR	SR
Steam locos per route mile				
	1.34	1.17	1.07	1.09
Coaching stock per route mile				
	3.47	3.03	2.66	4.99

	GWR	
Steam tender locos		1,550
Steam tank locos		2,500
Coaching stock		10,100
Frieght stock		90,000
Service stock		9,700
Electric motor & trailer stock		60
Electric locos		0
Railmotors		70

	SOUTHERN	
Steam tender locos		1,240
Steam tank locos		1,150
Coaching stock		10,800
Freight stock		7,500
Service stock		2,280
Electric motor & trailer stock		460
Electric locos		0
Railmotors		14

LONDON MIDLAND AND SCOTTISH

LONDON AND NORTH EASTERN

LONDON MIDLAND AND SCOTTISH

GREAT WESTERN

GREAT WESTERN

SOUTHERN

All statistics year ending 31/12/1921

The variety of locomotive, carriage and wagon stock inherited by the four main line companies was enormous. Locomotive building had continued right up to grouping, with the accumulated experience of war work to add refinement. Several companies had embarked upon 'big engine' projects, among them the North Eastern Railway under Sir Vincent Raven. The new mechanical engineers faced the task of rationalising the resulting array of types and designs, and that could not be done overnight. The GWR faced the least problems because it was blessed with considerable continuity: for the Great Western, the grouping largely rubber-stamped a process of amalgamation undergone years before. The remaining three companies, however, faced a daunting prospect. In some cases it was ten years before the goal was realised. Workshop rivalries died hard and there were difficulties in establishing the right form of structural organisation. But eventually each company developed its own engineering ethos, even if certain pre-grouping designs saw a restricted life as a result.

Plate 56: Southern Railway N15 on an express passenger service. The class was developed from a 1917 design of the SECR.

Plate 57: Sir Vincent Raven's Pacific design for the North Eastern Railway, seen here in LNER days.

On the European continent, the inter-war years saw considerable investment in electric and diesel traction on railways. In Britain the railways continued to maintain their faith in steam. The Southern Railway made rapid progress with electric traction in the 1930s, but by 1939 no more than five per cent of British railway mileage was electrified. Likewise, the Great Western introduced a fleet of diesel railcars for branch and cross-country operations, but found few imitators amongst its rivals. Why was there such a neglect of more efficient forms of traction in Britain? One possible explanation is that in the 1930s there was still too little known about the comparative costs and merits of different forms of traction. Moreover, it was also generally felt that the Southern Railway's traffic was more suited to

electric traction than that of the other companies, since the bulk of it consisted of passengers moving in dense flows over relatively short distances at peak times of the day. This type of traffic required rapid transit to avoid congestion and here the multiple electric unit, with its speed, acceleration and flexibility, scored heavily over the steam-hauled train. But perhaps the most important reasons for the lag in new technology were the capital locked up in steam traction which had yet to be proved obsolete, and the fact that generations of railwaymen — directors, traffic managers, engineers and workmen alike — had been reared in an era of steam traction and required much convincing to depart from established practice.

Steam Locomotive Renewals: 1921-38

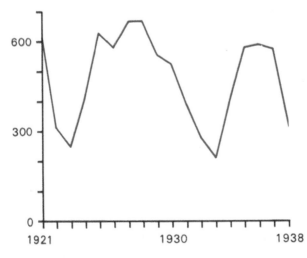

Plate 58: SR 'Lord Nelson' class on the Down 'Bournemouth Belle' approaching Surbiton. For a brief period this was the most powerful locomotive type in Britain.

Plate 59: 4-6-0 no. 6001 King Edward VII. The 'King' class represented the ultimate development of the Great Western's steam practice.

Plate 60: Gresley's experimental W1 4-6-4 express locomotive, equipped with a marine boiler and streamline casing. The novel external design was to become familiar on the LNER with the appearance of the P2 2-8-2s and the A4 Pacifics.

Plate 61: The first major locomotive design of the LMS — Fowler's 'Royal Scot' class. Here the engine of the same name is seen in charge of an express on the west coast main line.

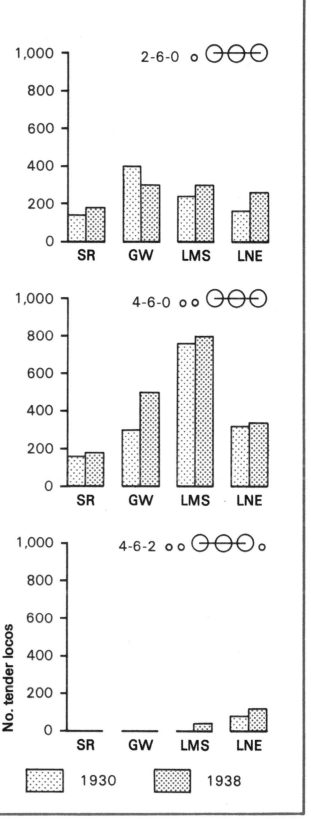

1938	Total steam locos	Tender	Tank	
SOUTHERN	1816	1082	734	
GWR	3630	1320	2310	
LMS	7613	5176	2437	
LNER	6518	4279	2239	

Committed to steam, the railway companies devoted much finance and energy to improving the performance of the steam locomotive. The LNER under Nigel Gresley became the initial pace-setter, but the other companies soon followed and by the late 1930s some fine locomotives were in service. Considerable effort went into the perfection of express locomotives, capable of handling heavy loads over long distances. The common wheel arrangements were 4-6-0 and 4-6-2, although Gresley provided the LNER with 2-6-2 and 2-8-2 designs as well, the latter for some of the difficult Scottish routes. The operational success of many of these improvements reinforced the view of railway managers that the steam locomotive had a future, a view no doubt confirmed by the remarkable performance of the LNER Pacific *Mallard* which, on test runs in 1938, attained a top speed of 126 mph, a British and world record for steam that has never been beaten.

A NEW AGE IN PASSENGER SERVICES

3860   3860

Increase in Daily Trains from London: 1918 to 1938

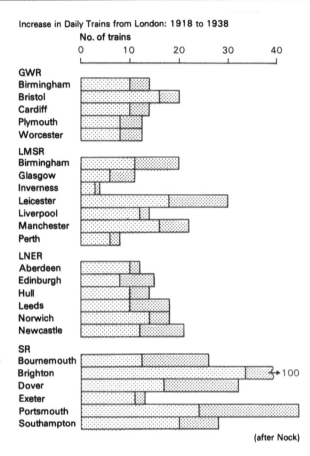

Between the wars, railways remained the undisputed leaders in long-distance passenger transport and the four companies made strenuous efforts to consolidate and publicise that position. With renewed and improved motive power, they were able both to expand the frequency of service and to advance average speeds of travel. As time passed, something of a competitive spirit developed between the companies. Direct service competition was obviously limited, but the pursuit of excellence in train operation was open to all. Much attention came to be focused on non-stop working, where the GWR had for many years held the distance record with its Paddington-Plymouth run. In 1927, however, the record passed to the LMSR, and in 1928 to the LNER with its King's Cross-Edinburgh service of nearly 393 miles.

Plate 62: LNER Up express near Grantham, hauled by a Gresley A1 Pacific.

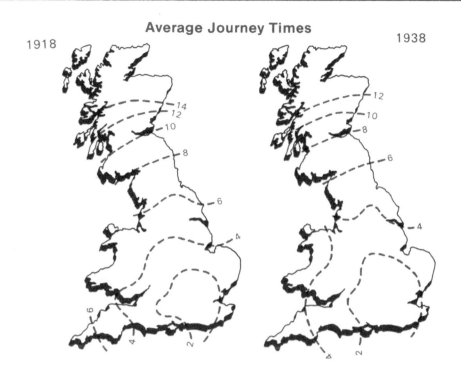

Average Journey Times

Plate 63: An example of the GWR's 'Castle' class — the outstanding workhorse of the company's passenger operations.

Plate 64: LNER 'Silver Jubilee' train near Abbots Ripton, circa 1936, hauled by A4 4-6-2 Silver Fox.

Plate 65: LMSR 'Coronation Scot' heading north over Bushey watertroughs near Watford in the summer of 1937. The locomotive is the first of the class, no. 6220 Coronation.

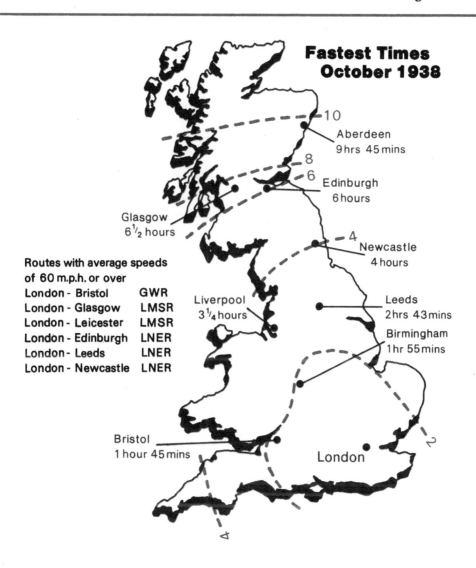

Fastest Times October 1938

10
Aberdeen
9 hrs 45 mins

8

6 Edinburgh
6 hours

Glasgow
6½ hours

Routes with average speeds
of 60 m.p.h. or over
London - Bristol   GWR
London - Glasgow   LMSR
London - Leicester   LMSR
London - Edinburgh LNER
London - Leeds   LNER
London - Newcastle LNER

Liverpool
3¼ hours

4 Newcastle
4 hours

Leeds
2 hrs 43 mins

Birmingham
1 hr 55 mins

Bristol
1 hour 45 mins

London

2

4

Prestige Trains

Improvements in timetabling were fundamental to maintaining the railways' competitive position in long-distance passenger movement, but the more perceptive of company officers soon realised that this would not necessarily restore the railways to their former prominent place in public imagination. Hence there began a campaign to create a dynamic and futuristic image, a search for what today would be called 'psychic income'. It was Gresley who led the field once more. In 1935 he introduced on the LNER's east coast main line a prestige train appropriately named the 'Silver Jubilee'. Both locomotive and carriages were extensively streamlined, following the

fashion of the age, and full use was made of modern materials like stainless steel and rexine in surface finishes. Not surprisingly, the train was operated to the fastest schedule then in use. So successful was the enterprise that Gresley followed it with further high-speed streamliners in 1937, including the famous 'Coronation' train with its beaver-tail observation coach and six-hour timing to Edinburgh. As the LNER's competitor for Anglo-Scottish traffic, the LMSR followed Gresley's example by introducing in 1937 its own streamliner, the 'Coronation Scot', to operate between Euston and Glasgow in six and a half hours.

SOUTHERN ELECTRIC

3860   3860

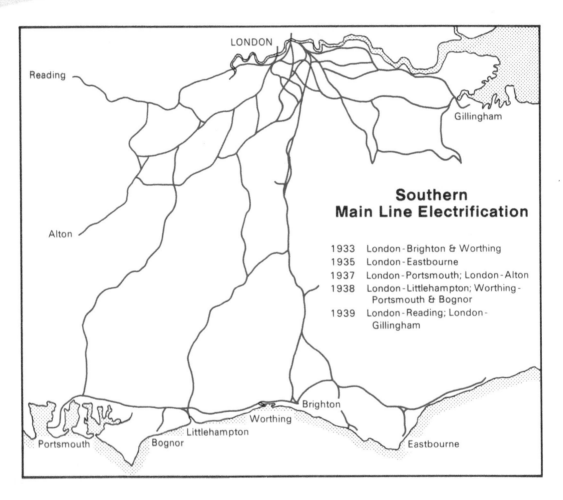

**Southern
Main Line Electrification**

1933	London - Brighton & Worthing
1935	London - Eastbourne
1937	London - Portsmouth; London - Alton
1938	London - Littlehampton; Worthing - Portsmouth & Bognor
1939	London - Reading; London - Gillingham

The Southern Railway's electrified system had no counterpart in Britain in 1939. Early in 1938 the GWR had commissioned an initial survey for the electrification of their lines west of Taunton as a possible answer to the high cost of supplying loco coal to the area. However, the projected investment return was less than one per cent and no action was taken. Electrification had been introduced on some London suburban lines well before the formation of the Southern Railway Company.

Electric Railways in Britain by 1939
Route miles

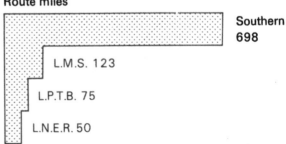

Southern 698

L.M.S. 123

L.P.T.B. 75

L.N.E.R. 50

But under the enterprising management of Sir Herbert Walker, the Southern embarked upon a major scheme of main line electrification, using a third-rail DC system. By 1939 just over 30 per cent of the company's route mileage had been converted, and electric services accounted for no less than 60 per cent of the train mileage operated. The Southern electrics became renowned for their reliability and frequency, and in every case traffic increased. During the 1930s passenger journeys on electrified lines rose by 12.5 million, compared with a drop of one million on the remaining steam services. Operating costs were one-third to one-half less than those of steam, the main savings being on labour and maintenance costs.

Plate 66: SR electric multiple units form a stopping service to Portsmouth in July 1937 and are seen near Farncombe.

Plate 67: Steam-hauled fast train passing through Box Hill on the Mid-Sussex line of the former LBSCR. The train includes a solitary Pullman carriage.

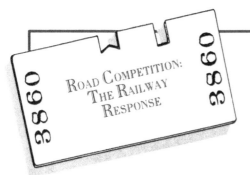

Company Road Vehicles: 1938

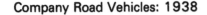

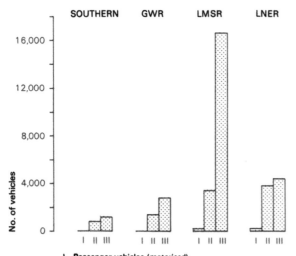

I   Passenger vehicles (motorised)
II Parcels and goods vehicles (motorised)
III Parcels and goods vehicles (horse-drawn)

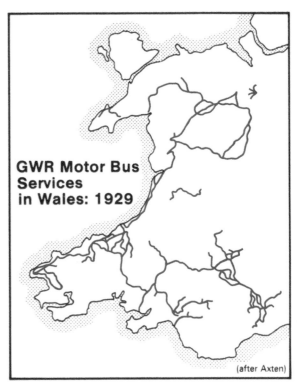

GWR Motor Bus Services in Wales: 1929

(after Axten)

Throughout the inter-war years the railways faced a growing problem of competition from road transport. Traffic demands were growing fastest in the Midlands and the South, but the railways faced great difficulties in adapting their operations away from the industrial heartlands of South Wales, Scotland and the north of England. The road transport industry was quick to capitalise on the gap and by September 1938 there were half a million road goods vehicles in use, 53,000 buses and coaches, as well as nearly two million private cars. Road transport also offered a more flexible and convenient service, which was often cheaper as well. As time passed, therefore, the railways not only lost existing traffic but failed to gain new customers. In an attempt to salvage their position, the main line companies entered road operations themselves, especially after full statutory powers were granted in 1928. To begin with they ran their own fleets, but later the emphasis was placed on acquiring financial interests in established enterprises. The railways seem to have gained little from this move into road transport. The capital return was poor and the scale of their interests was too small to curb competition significantly.

Plate 68: LMS 5-ton lorry by AEC.

Plate 69: LNER bus, built in 1928 for Durham and Newcastle country services.

RAILWAY AIR
SERVICES

3860   3860

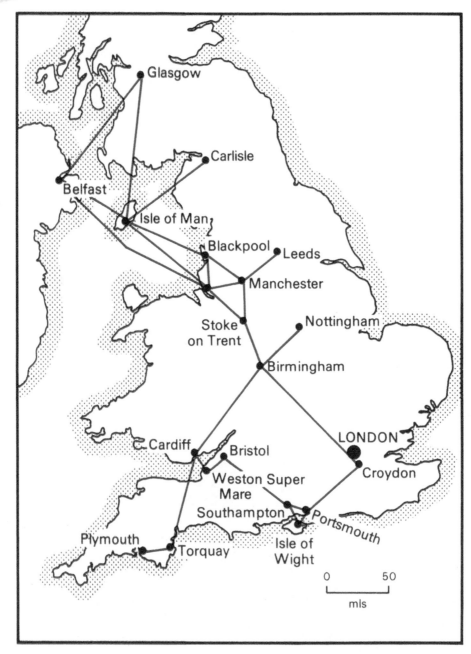

The principal routes of Railway Air Services Ltd, Summer 1936

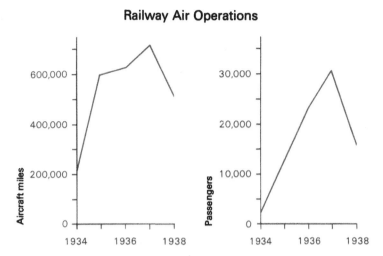

Railway Air Operations

The pioneer internal air services of the inter-war years were never a competitive threat in the way that road services were. However, with a rather unusual display of foresight, the railway companies perceived a threat in the future, especially over the longer routes. If they could participate themselves before a strong foothold had been established, the experience of road transport might be avoided.

In the 1930s, therefore, the railways acquired a controlling interest in most of the long-distance air services operating within Britain. Financially the venture was unrewarding and traffic was limited and variable. But the progressive image to which it contributed was not unimportant and there were some remarkable instances of integrated timetabling.

Plate 70: A four-engined aircraft belonging to Railway Air Services Ltd.

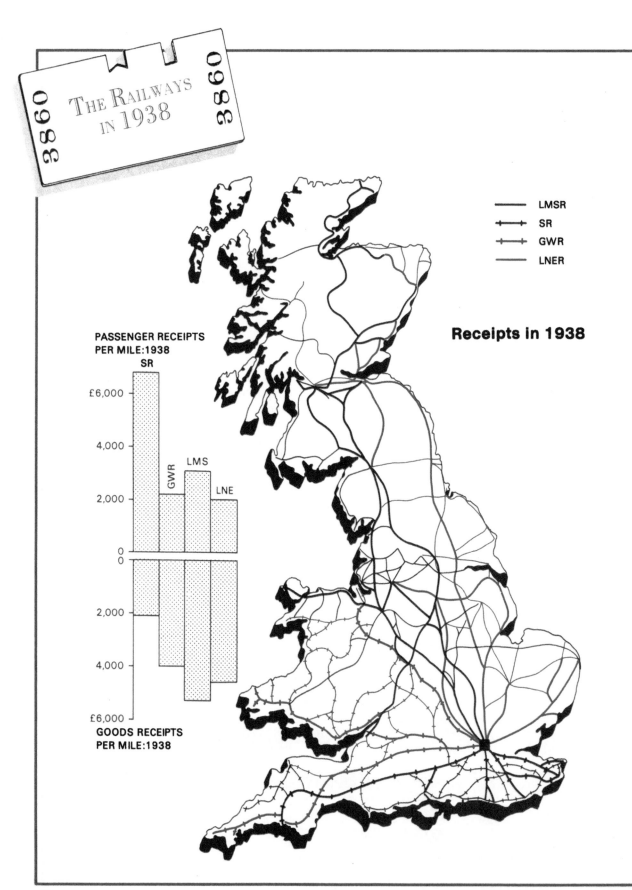

THE RAILWAYS IN 1938

3860

Receipts in 1938

LMSR
SR
GWR
LNER

PASSENGER RECEIPTS
PER MILE:1938

SR

£6,000

4,000

GWR

LMS

2,000

LNE

0

0

2,000

4,000

£6,000

GOODS RECEIPTS
PER MILE:1938

Passenger Operations

For all their worth, the accelerated main line steam services and the streamliners did comparatively little to lift the overall profitability of passenger operations between the wars. The LNER and the GWR, with their elaborate webs of rural lines in areas like south-west England and East Anglia, were thwarted from the beginning. Their passenger receipts per mile were a third of those returned on the Southern in 1938. Even in the Victorian era, the number of country services which made a profit was minimal. Faced with road competition, the cause was a lost one. Between the wars, 240 miles of track and 350 stations were closed to all traffic, and 1,000 miles of track and 380 stations closed to passengers but retained for goods traffic only. But much of what remained was still of questionable economic viability. It was left to the Southern to show what could be achieved, although the compactness of its territory and the characteristics of its traffic provided rather an unfair comparison.

Plate 71: Modernity in passenger train technology was parallelled by developments in railway architecture. This is the Southern's new station at Malden Manor, opened in 1938.

Plate 72: One of the LMS Beyer-Garratts heads an Up coal train through Elstree.

Plate 73: 12-ton private owner wagon.

Freight Traffic: 1938 (million tons)

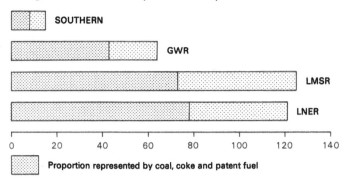

Freight Vehicle Capacity: 1938

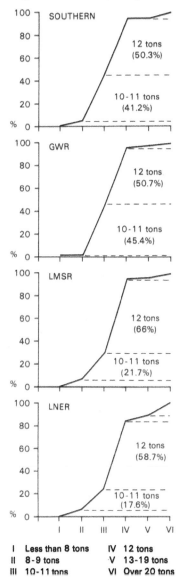

I   Less than 8 tons   IV   12 tons
II   8-9 tons   V   13-19 tons
III 10-11 tons   VI   Over 20 tons

Plate 74: GWR 20-ton wagon.

The Freight Sector

In the years after the First World War freight traffic fell away consistently. In no single year did the volume of traffic attain the 1913 level, and all classes of freight — merchandise, coal and minerals — experienced a decline. Even in the best year of 1937 the total volume of freight traffic was some 18.4 per cent lower than in 1913. The LMS had the best per-mile receipts, undoubtedly benefiting from its occupation of the new industrial heartlands of the Midlands.

A persistent problem with freight operations had always been the small average size of wagons, carrying between 10 and 12 tons. In pre-grouping days, the North Eastern, amongst others, had introduced 20-ton vehicles and one would have expected a continuation of this trend by their successors as the search for economies was made. As the graphs reveal, however, 10-12 tons remained the dominant vehicle size in 1938. Only the LNER had made any measurable progress.

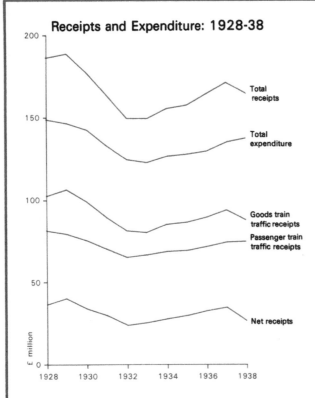

Receipts and Expenditure: 1928-38

Total receipts

Total expenditure

Goods train traffic receipts

Passenger train traffic receipts

Net receipts

£ million

1928 1930 1932 1934 1936 1938

The railways' financial position was made worse by the effects of the world economic slump after 1929. Receipts fell drastically in the early 1930s, seriously affecting the ability to invest. There was a gradual recovery from 1933 but the damage proved irreparable. Under a more stable economic climate the four companies might have responded to competition with greater success than they did, but with so much of their revenue dependant on a buoyant export market, they were largely lost.

The one alternative form of transport in which the railways met success was the short-sea crossing. The seeds of this had been sown early and even by 1914 the London and South Western and the Great Central had become leaders in the field, sometimes also constructing elaborate dock systems for more general usage. The main line companies inherited the full range of these interests and continued to exploit them up to 1939.

Plate 75: Diesel railcar no. 4 — the Great Western's bold attempt to look into the future.

Plate 76: The SR's Isle of Sark *was brought into service in 1932 for the Channel Islands route.*

Plate 77: The LMS on the Clyde: paddle steamer arriving at Gourock Pier.

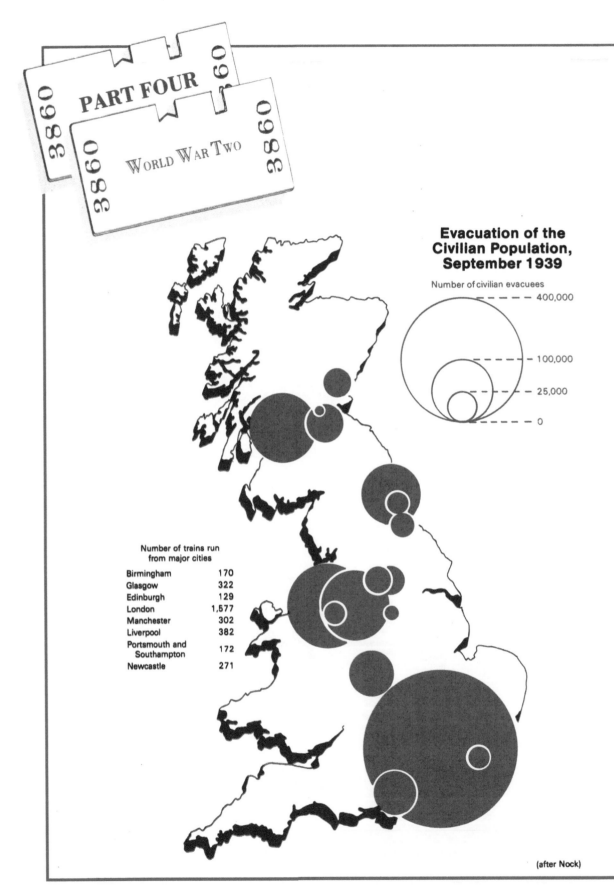

Evacuation of the Civilian Population, September 1939

Number of civilian evacuees

- 400,000
- 100,000
- 25,000
- 0

Number of trains run
from major cities

Birmingham	170
Glasgow	322
Edinburgh	129
London	1,577
Manchester	302
Liverpool	382
Portsmouth and Southampton	172
Newcastle	271

(after Nock)

During the Second World War the railways' capacity was utilised almost to breaking point, to satisfy heavy traffic demands over longer distances. The virtual suspension of private motoring meant that the public transport system bore the brunt of the demands for conveyance, which included the large scale evacuation of people from cities and the enormous requirements arising from military operations. During hostilities the railways ran no less than 538,559 special trains on behalf of the government. Total passenger movement almost doubled between 1938 and 1945, while freight traffic rose by 50 per cent between 1938 and the peak year of 1944. Moreover, the railways accomplished this task without any increase in their physical assets, which if anything tended to decline owing to an increasing amount of rolling stock under repair. It was achieved therefore by a much more intensive and efficient use of existing resources.

Following the declaration of war, the railways were placed under government control and the companies were guaranteed a net income of £43m per annum. By the end of the war, however, the railways faced grave problems as a result of bomb damage, the enforced reduction in maintenance levels and investment, and the intensive use of stock. Physical loss due to aerial warfare has been estimated at more than £30m, while arrears of maintenance amounted to at least £150m.

FREIGHT TRAFFIC			
Net ton-miles (all railways) – in millions –			
	Pre-war	1943	
Merchandise	5,192	9,659	86% increase
Minerals	3,182	5,355	68% "
Coal, coke & patent fuels	8,295	9,343	12.5% "
	16,669	24,357	(after Nock)

Passengers Per Train-Mile

GWR
LMS
LNER
SOUTHERN

☐ Pre-war
▨ 1944

Plate 78: Bomb damage at St Pancras station during the London 'blitz'.

Plate 79: The LNER's late entry in the electrification stakes: the Manchester-Sheffield-Wath scheme. Here the prototype locomotive is seen on its trial run in October 1941.

Plate 80: O.V.S. Bulleid's wartime design for the Southern Railway was a streamlined Pacific. Here 'Merchant Navy' class no. 21C15 Rotterdam Lloyd leaves Waterloo with the inaugural 'Devon Belle' in the summer of 1947.

The effects of war conditions were reflected in the reduced capacity of the railways in the immediate post-war years. A substantial proportion of the permanent way and rolling stock was then awaiting repair, in some cases up to 30 per cent. The difficult economic conditions then prevailing, including severe shortages of raw materials and labour, meant that much of the repair work was delayed. In addition, a lot of the rolling stock was antiquated anyway; for example, some 50 per cent of the freight wagons were 35 years old and at least 10 per cent were ready for scrapping. Much the same could be said for passenger rolling stock. Only the locomotive stock was up to reasonable strength initially, though many locomotives were either badly in need of repair or had reached the end of their useful working life. Sir Nigel Gresley's untimely death in 1941 and Sir William Stanier's move to the Civil Service in 1944 had removed formidable talents from the field of British locomotive design. O.V.S. Bulleid, Gresley's former assistant, nurtured something of the same spirit on the Southern, but conditions in the immediate post-war years were unpropitious for experiment and novelty in design.

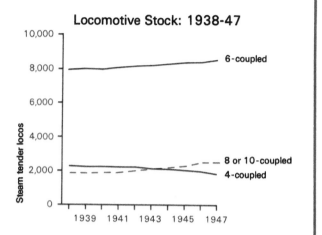

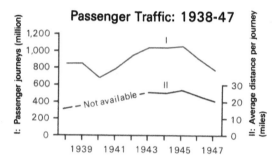

Plate 81: Bulleid's ill-fated 'Leader' class never saw regular service on the SR. The prototype, no. 36001, is seen at Eastleigh in 1950.

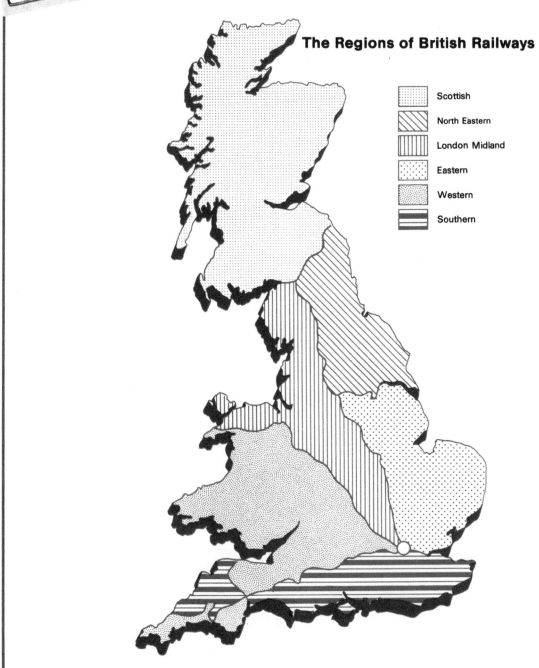

The Regions of British Railways

▦	Scottish
▨	North Eastern
▥	London Midland
▦	Eastern
▦	Western
▤	Southern

The election of a Labour Government in the summer of 1945 effectively sounded the end for private railway operation in Britain. State control of key sectors of the economy, transport especially, had long been a central plank of Labour policy and thus under the 1947 Transport Act British Railways was born. Overall control passed to the British Transport Commission which was charged with the task of achieving an efficient, economical and integrated system of public inland transport. Actual management of the railways passed to the Railway Executive, working within guidelines set by the Commission. For administrative purposes the Executive divided BR into six regions. To some extent these matched the old company organisation, but in the creation of a North Eastern Region the Executive was reverting to a pre-grouping arrangement, while the Scottish Region recalled an alternative 'grouping' scheme of 1920 in which the railways of Scotland were to have been operated separately.

THE MAIN INHERITANCE		
(1946)		
London Midland and Scottish	Route miles	6,785
	Capital issued	£413,778,857
	Net revenue	£15,923,680
London and North Eastern	Route miles	6,333
	Capital issued	£376,483,643
	Net revenue	£11,078,471
Great Western	Route miles	3,741
	Capital issued	£149,773,213
	Net revenue	£7,467,390
Southern	Route miles	2,156
	Capital issued	£169,290,003
	Net revenue	£7,184,536

Plate 82: H.G. Ivatt's two main line diesel-electrics were built for the LMS but are seen here in BR ownership double-heading the 'Royal Scot' near Tring.

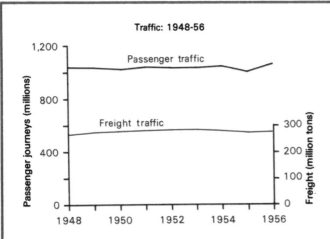

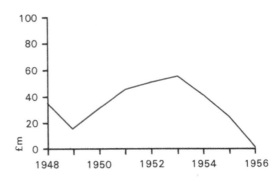

The British Transport Commission inherited a railway system which was in very poor physical shape. There was an enormous backlog of investment to be made good which would take several years to complete, while the structural and administrative reorganisation following nationalisation proved a more difficult and protracted task than at first envisaged. Hence in the first few years the BTC was preoccupied with reorganising the system and maintaining it in reasonable working order rather than projecting plans for the future. Fortunately, traffic volumes in these years remained well up on pre-war times as the threat of road competition was restrained by austerity restrictions, notably petrol rationing. Thus until the mid-1950s the railways continued to break even on their operating account. Nevertheless, against the halcyon days of the later thirties, British Railways struck a shadowy image in its first years. The streamliners made no reappearance. Train schedules fell well short of those sustained before the war. General standards of presentation often left much to be desired. An attempt was made to develop a corporate image by introducing new locomotive and passenger stock liveries and new schemes for station colours, but the results were not very successful. No single colour scheme was employed, for instance, so that the outcome was disparity not unity.

Plate 83: The 'Tees-Tyne Pullman', hauled by ex-LNER A4 Pacific Silver Fox, *in early BR days.*

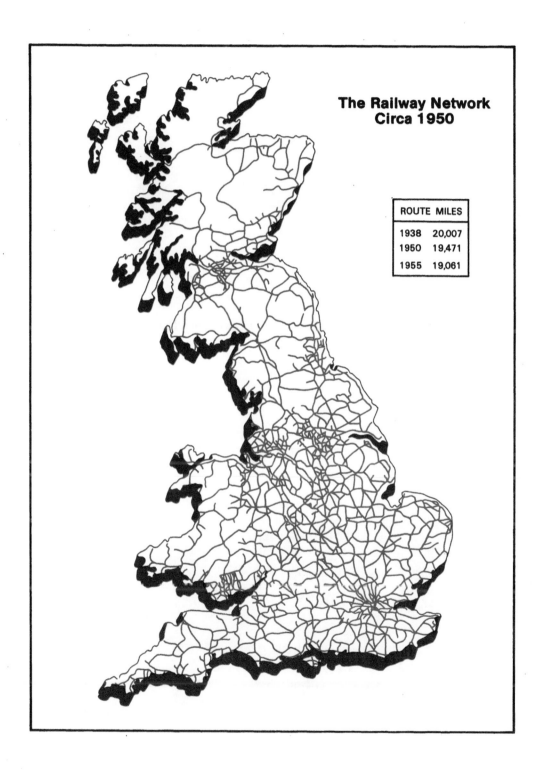

**The Railway Network
Circa 1950**

ROUTE	MILES
1938	20,007
1950	19,471
1955	19,061

Plate 84: BR Standard class 4 no. 75006 takes an Up train away from Didcot in 1951.

Plate 85: The Ostend Boat Express, in the charge of BR Standard class 7 no. 70014 Iron Duke, *is seen in 1952.*

Plate 86: Locomotive exchanges, 1948: ex-LMS Pacific no. 46236 City of Bradford *ventures onto the Western Region with a Plymouth-Paddington express.*

One of the main benefits which nationalisation was expected to bring was the ability to standardise stock and equipment and thereby achieve substantial operating economies. For a few years after 1948, railway workshops continued to turn out locomotive designs of the former companies, but soon plans were laid for a series of new, universal designs. Existing steam practice was carefully examined in a comprehensive series of interchange trials in which selected locomotives were tested outside their normal operating areas. The outcome was a major programme of steam locomotive building embracing almost every sphere of railway operation. Much argument has centred upon

the decision to re-equip with steam rather than some alternative traction. The BTC favoured a full-scale investigation of diesel and electric traction. The Railway Executive, with many devotees of steam amongst its number, thought otherwise. In some respects the dilemma was similar to the one faced between the wars. When the need for new motive power was pressing, it was safer and easier to opt for adapting established practice than to back a relatively untried alternative.

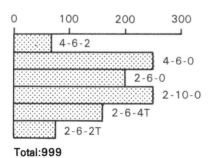

BR Standard Steam Locomotives

Total: 999

New Locomotive Stock: 1948-56

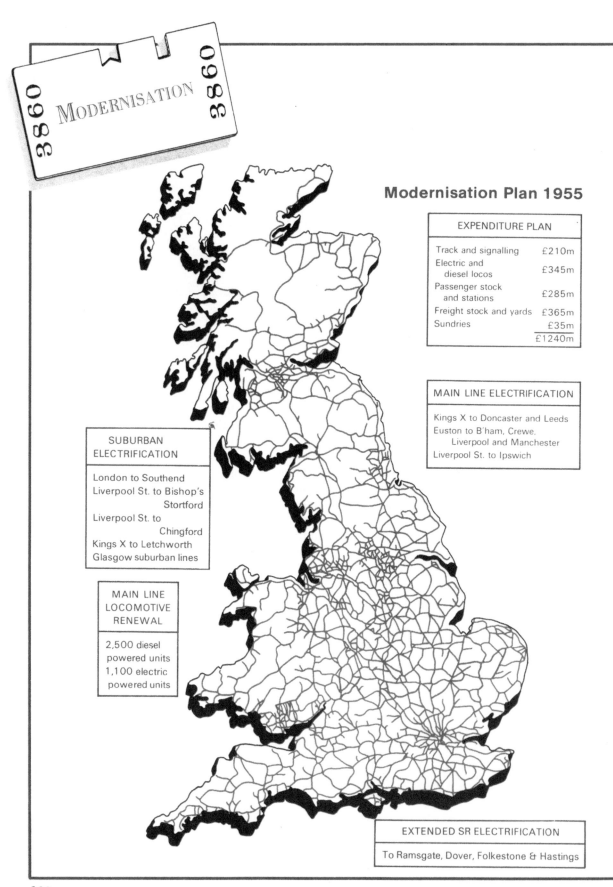

3860
3860 MODERNISATION **3860**

Modernisation Plan 1955

EXPENDITURE PLAN	
Track and signalling	£210m
Electric and diesel locos	£345m
Passenger stock and stations	£285m
Freight stock and yards	£365m
Sundries	£35m
	£1240m

MAIN LINE ELECTRIFICATION

Kings X to Doncaster and Leeds
Euston to B'ham, Crewe,
   Liverpool and Manchester
Liverpool St. to Ipswich

SUBURBAN ELECTRIFICATION

London to Southend
Liverpool St. to Bishop's
   Stortford
Liverpool St. to
   Chingford
Kings X to Letchworth
Glasgow suburban lines

MAIN LINE LOCOMOTIVE RENEWAL

2,500 diesel
powered units
1,100 electric
powered units

EXTENDED SR ELECTRIFICATION

To Ramsgate, Dover, Folkestone & Hastings

Plate 87: BR class 42 'Warship' diesel-hydraulic on the 'Cornish Riviera' train in 1960.

By the mid-1950s, signs of long-term decline were becoming apparent for Britain's railways. As the national economy picked up after years of austerity, it was road rather than rail transport which attracted new traffic. The industrial trends initiated between the wars now gathered pace so that a steadily lessening proportion of internal trade demand was favourable to rail. It was also becoming clear that large parts of the railway network were entirely uneconomic, and yet little progress was made in excising the parts that did not pay. Between 1948 and 1953, for example, only 253 miles were closed to all traffic. The railways' fundamental problem, however, was that their cost-price structure remained adverse because of government control on charging at a time when costs were rising rapidly.

Plate 88: BR two-car lightweight diesel multiple unit built in 1954 for service on the North Eastern Region.

In an attempt to check this decline and in a determined effort to carry the railways into a new era, the BTC launched its Modernisation Plan for the railways in 1955. This involved an ambitious programme of re-equipment at an estimated cost of £1,240m — later revised to £1,660m — and the whole scheme was to be completed within fifteen years. It was anticipated that the economic benefits accruing would allow the railways to break even.

The Modernisation Plan of 1955 has been severely criticised on the grounds that it was little more than a last-minute rescue operation, hastily conceived, ill-thought out and badly costed, which the Ministry of Transport and the government accepted largely because they had nothing better to offer. The most glaring fault was the failure to do any comprehensive costing of the London Midland electrification proposal. Despite these drawbacks, much of the plan went ahead fairly rapidly until it was brought to a halt in 1961 pending further changes in management and planning. Between 1955 and 1962 railway investment was running at three times the level of 1948-54 and the benefits of modernisation soon became apparent to the traveller. Perhaps the most visible was the rapid conversion to diesel traction. New construction of steam locomotives ceased in 1960, and by 1963 diesel and electric power accounted for some 62 per cent of total traction miles run, compared with 13 per cent in 1955. Some 4,000 separate diesel multiple units were introduced and various sections of the network were electrified, including a start on the major

Plate 90: BR Type 2 no. D5082 takes charge of the 'Condor' express container service in 1959.

London Midland scheme. Comprehensive dieselisation had a very obvious impact upon passenger train schedules, especially on the main lines. On the east coast Anglo-Scottish route the powerful 'Deltic' locos had by 1962 re-established the six-hour King's Cross-Edinburgh schedule achieved in 1937 by the LNER 'Coronation' and with a much heavier load. The operating economies were such that the total fleet of 22 Deltics was able to perform duties which would previously have required some 55 steam locomotives. Compared with passenger operation, the achievements of the Modernisation Plan on the freight side were unspectacular. Steam traction

Plate 89: The prototype 'Deltic' locomotive approaches Preston. This was the forerunner of the class 55 diesel-electrics which dominated passenger operations on the east coast main line for twenty years.

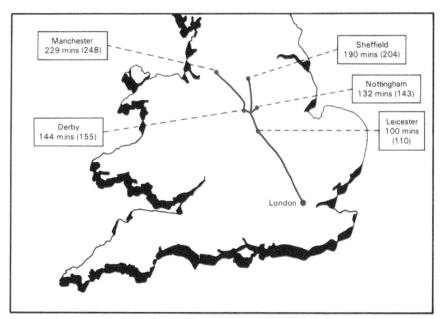

Re-scheduling of passenger services on LMR Midland lines, Winter 1962-3. Average times are given, with previous journey times in brackets.

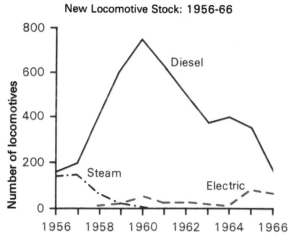

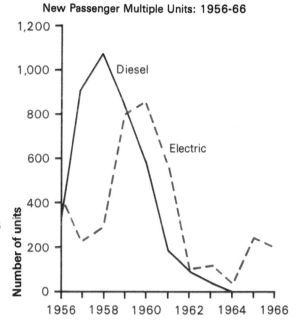

still accounted for over half the traction miles in 1964 as against 13.2 per cent for passenger traffic. Some progress was made with rationalising and modernising out-of-date freight services. Freight rolling stock was reduced substantially, larger wagons were introduced, freight trains were speeded up and traffic was concentrated at fewer and more modern goods terminals and marshalling yards. Priority was given to the development of high-speed movement of bulk freight in specially designed wagons. Containerisation of loads was encouraged and an early example of specialised container train operation was the 'Condor' express service between London and Glasgow, which started in 1959.

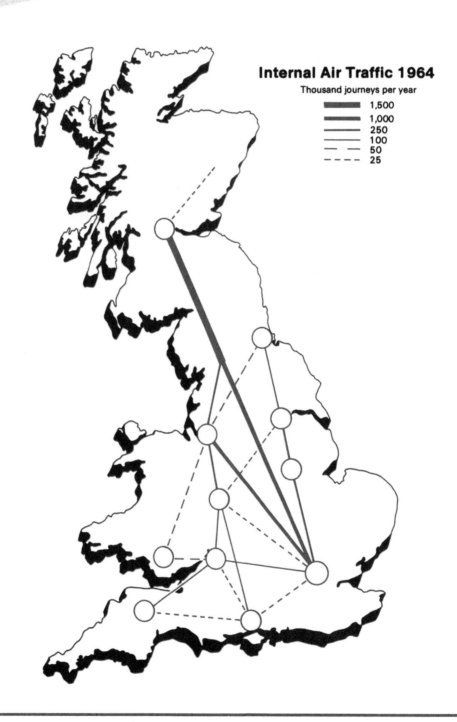

Internal Air Traffic 1964

Thousand journeys per year

▬▬▬	1,500
▬▬	1,000
───	250
───	100
─ ─ ─	50
- - -	25

Percentage Shares of Goods Traffic

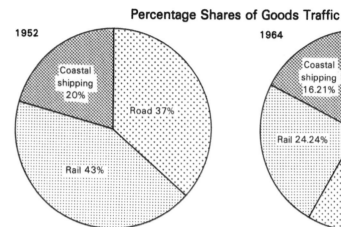

1952

Coastal shipping 20%

Road 37%

Rail 43%

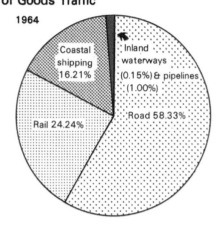

1964

Coastal shipping 16.21%

Inland waterways (0.15%) & pipelines (1.00%)

Road 58.33%

Rail 24.24%

From 1955 the railways faced severe and growing competition from road transport. In 1953, the year petrol rationing ceased, there were 2.75 million private cars in Britain. This compared with approximately two million at the outbreak of war. Within ten years, however, the total had increased dramatically to some seven million. At first the competitive effect was concentrated on shorter-distance movement, but with trunk road improvement and the development of motorways — the first stage of the M1 opened in 1959 — long-distance travel became involved. Bus operators were quick to exploit the speed advantages of the motorways and established express services between cities which were in direct competition with the railways' own express passenger operations. The buses could not quite match the railways in journey times, but they were able to undercut their fares substantially. As road haulage was made easier and quicker, so more and more freight traffic was attracted. The advantages of door-to-door delivery and the ability of firms to set up their own lorry fleets were critical supplementary factors.

Towards the end of the decade, the railways also faced competition from domestic airways. By 1963 British European Airways were running five return trips daily between London and Glasgow. The travel time was only one and a half hours and fares bore comparison with the first-class sleeper accommodation of the railway route. Just as the railways were endeavouring to overcome problems of structural reorganisation and technical backwardness, they faced the most serious competitive threats of their existence.

Plate 91: The picturesque ex-Midland and South Western Junction line from Andover to Cheltenham was an early victim of road competition.

The results of the Beeching survey of freight traffic. From The Reshaping of British Railways, Part I. *Reproduced by permission of the Controller of Her Majesty's Stationery Office.*

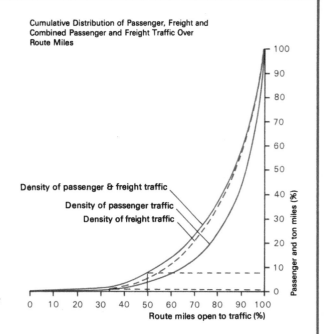

Cumulative Distribution of Passenger, Freight and Combined Passenger and Freight Traffic Over Route Miles

Density of passenger & freight traffic

Density of passenger traffic

Density of freight traffic

Passenger and ton miles (%)

Route miles open to traffic (%)

Beeching: the tell-tale graph which revealed the uneconomic state of much railway operation. In passenger movement, one-third of the route mileage carried only one per cent of traffic.

Despite the visible benefits of modernisation under the 1955 Plan much remained to be done when the programme was suspended in the early 1960s. At this time the government was concerned that modernisation had done little to check the declining fortunes of the railways. Annual deficits were rising at an alarming rate during the late 1950s, made worse by the escalating competition from road transport, especially on the freight side.

In an effort to arrest the decline, policy was switched to a more commercial approach. In May 1961 Dr Richard Beeching, the former Chairman of Imperial Chemical Industries, was appointed to manage the railways and given a mandate to make them pay. Public transport was reorganised under the Transport Act of 1962, involving the abolition of the BTC and the creation of separate boards for each branch of transport. The new British Railways Board was given considerable commercial freedom, and a complete financial reconstruction was carried out to relieve the financial burden of the new Board. As a result of these changes the emphasis shifted from

re-equipment to closures and rationalisation.

Beeching's first task was to carry out a thorough survey of the network to find out which services paid their way and to determine exactly where the losses were being made. The results of this exercise were published in March 1963 in *The Reshaping of British Railways*. This was in fact the first detailed cost study of the whole railway system and it confirmed the suspicions of those who maintained that the greater part of the railway network was uneconomic. Much of the system was found to be under-utilised. Indeed, traffic density was so low on half the system that it was barely sufficient to cover the cost of providing the route (track and signalling), and allowed nothing for movement and other costs. Conversely, the remaining half of the system earned enough to cover its route costs more than six times. The position was similar in the case of stations and rolling stock. One half of the stations produced only two per cent of the passenger receipts, while one half of the freight wagons were surplus to requirements.

The Beeching Plan:
Passenger Services
to be Withdrawn

Plate 92: Ex-LMS 2P 4-4-0 no. 40634 pilots a BR Standard class 4 on the 'Pines Express' near Midford in 1960, on the picturesque Somerset and Dorset Joint line — one of many routes closed under the Beeching 'axe'.

Dr Beeching's proposals for reform were drastic. A large part of the network would have to be abandoned, including the closure of 5,000 route miles and over 2,000 passenger stations, as well as the withdrawal or modification of 400 passenger services. The second proposal was for selected development and rationalisation of the key inter-city routes — the ones which had proved most profitable in the survey of the network. Details of this scheme were published separately in February 1965. Of the total 17,000 route miles, it was envisaged that only 8,000 miles would ultimately be retained, of which about 3,000 miles were selected for intensive development. The other proposals were concerned largely with the modernisation and reform of rolling stock and freight traffic. The amount of rolling stock was to be reduced substantially and modernised, and steam traction phased out completely. A radical overhaul of freight handling methods was planned, including the introduction of a new type of service — the liner train — to provide combined road and rail movement of containerised merchandise.

Plate 93: Graveyard for Great Western steam: Swindon locomotive dump in 1961.

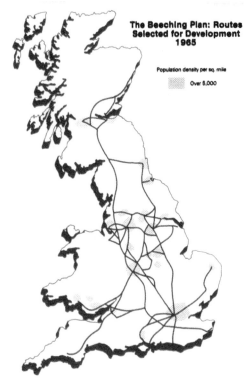

The Beeching Plan: Routes Selected for Development 1965

Population density per sq. mile

Over 6,000

Most of the Beeching proposals were implemented quickly, though the pace of progress was slowed in the late 1960s pending further changes of policy. By 1970 the railways were much slimmer than they had been a decade before, though financially they were still destitute. There had been a massive reduction in assets. Between 1962 and 1970 the number of locomotives was reduced by nearly two-thirds, passenger coaches by 45 per cent, marshalling yards by three-quarters, route mileage by one-third and employees by nearly one-half. There was, however, a more constructive side to the Board's work. The phasing out of steam traction was completed in 1968, while main line passenger services were improved in terms of speed and comfort. The most spectacular achievement was the completion of the London Midland

Plate 94: Electric haulage for a 'freightliner' train on the London Midland Region in 1966.

electrification which cut journey times
between London and north-west England by
about one-third. Freight handling methods
were also substantially improved; the trainload
rather than the wagonload became the main
unit of movement as a result of the
introduction of company trains, the
merry-go-round train for coal and the
freightliner. At the same time, further progress
was made in the programme of renewal of
track, structures and signalling which had been
started in the 1950s.

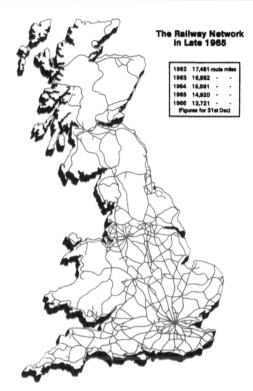

The Railway Network
In Late 1965

1962	17,481 route miles
1963	16,982 · ·
1964	15,991 · ·
1965	14,920 · ·
1966	13,721 · ·
(Figures for 31st Dec)	

Plate 95: The new Tyne Marshalling Yard, 1963.

Towards the end of the 1960s it was becoming all too clear that the railways would never pay their way. The Beeching reforms had managed only to stabilise the deficit. They were unable to reduce it in the face of continuing penetration of the market by road transport and the tendency for increasing costs to eat into economies as soon as they had been made. The Labour Government of the day was also concerned by the rapidly shrinking size of the network and therefore rejected commercial viability as the prime objective of railway policy. The railway system, it considered, should be designed to accord more closely with the country's economic and social needs. Accordingly it was decided to stabilise the future network at around 11,000 route miles. Under the Transport Act of 1968 the railways received massive financial assistance including a major capital reconstruction and track renewal grants. The legislation also recognised the need to subsidise uneconomic but socially necessary services by specific grants.

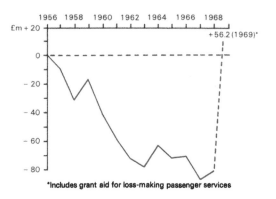

Net Revenue: 1956-69

*Includes grant aid for loss-making passenger services

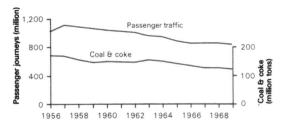

Traffic: 1956-69

Plate 96: An LMR passenger express approaching Cheddington in 1979, with an AC electric of class 86 at its head.

Plate 97: A High Speed Train of class 253 prepares to depart from Reading for Paddington as a westbound HST service draws in.

Britain's railways since 1970 have been characterised by an uneasy combination of failure and success. On the one hand, for example, the costs of stabilising the network and of underwriting increasing deficits in operations have presented an enormous burden to the Exchequer; by the early 1980s the annual subsidy was approaching £900m. On the other hand, the period has been one of remarkable achievement in inter-city

passenger operations. The introduction of the High Speed Trains (HSTs) from the mid-1970s has captured public imagination in a similar way to the streamliners of the 1930s. The distinctive wedge-shaped cab front of the HST has provided a compelling advertising image, while its smooth running and high operational speed (up to 125 mph) have turned an advanced image into concrete reality.

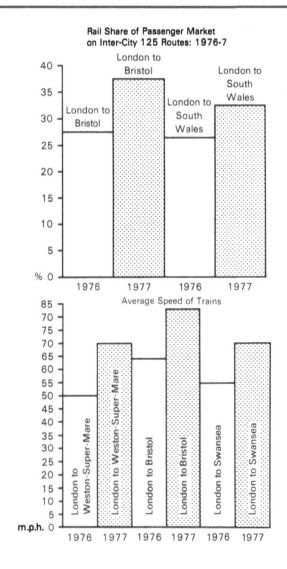

Rail Share of Passenger Market on Inter-City 125 Routes: 1976-7

Average Speed of Trains

Inter-city passenger movement has been the flagship of British Rail's operations in the late 1970s and early 1980s. Full electrification of the Euston-Glasgow main line brought the schedule of the 'Royal Scot' down to within five hours by the mid-1970s. In 1976 the first of the HSTs made their appearance and quickly won back custom to the railways with their superior standards of service. In the first full year of operation of the 125 mph HST services between London and Bristol/Weston-super-Mare and London and South Wales, the rail share of the passenger market on these routes advanced by 30 per cent. HST services have since been extended to other main lines, including the Anglo-Scottish east coast route. Between 1975 and 1980 the number of HST units cars and carriages rose from 10 to 772.

In its remaining passenger sectors, BR has faced growing problems of ageing stock and deferred maintenance of permanent way. In 1977 about one-third of electric multiple units were between 16 and 20 years old and a further quarter were over 21 years old. The position with diesel units was even worse, and in neither case was any major replacement of stock feasible within prevailing investment levels. On Tyneside the local metropolitan authority has adapted BR's former suburban operations and produced a highly successful system of rapid, light urban transit. Whether this provides a system that can be imitated elsewhere remains to be seen. But the case for investment remains, the more so when it is appreciated that the rail share of the passenger market had declined from 21 per cent in 1952 to 7 per cent in the early 1980s.

Plate 98: The Advanced Passenger Train makes a test run on the west coast main line near Beattock, 1980.

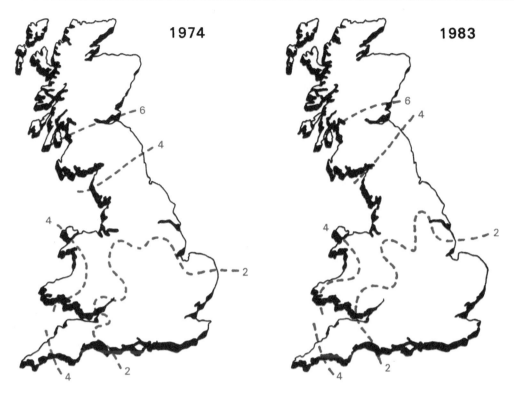

The impact of the High Speed Trains on journey times is revealed in this diagram of comparative two-hour isochrones from London in 1974 and 1983 (fastest timings).

Plate 99: Increasing traffic to London's second airport encouraged the introduction of a high-speed 'Gatwick Express' service in May 1984. Electro-diesel no. 73123 Gatwick Express heads away from Balham later that year.

Freight operations during the 1970s had little of the success recorded in passenger business. The contraction in the country's coal mining and steel making industries removed profitable freight custom, while the general deterioration in economic conditions from 1974 added to decline. By the early 1980s, freight carryings were about half the level recorded 30 years earlier, and the position is increasingly such that BR moves only the items which road transport is incapable of carrying. However, the freight business has not been without operational improvement. By 1977, for instance, 80 per cent of rail freight moved in trainloads rather than wagonloads. Nevertheless, the same year saw a reaffirmation of the value of wagonload traffic with the establishment of Speedlink. Using high-capacity air-braked wagons, capable of running at up to 75 mph between main industrial and market areas, it hopes to attract smaller businesses back to the railways.

Plate 100: A BR class 33 diesel-electric provides the motive power for a trainload of ballast from the Southern Region's quarry at Meldon.

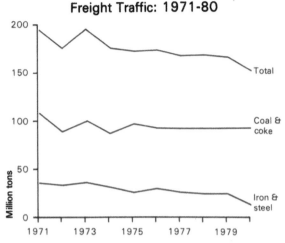

Freight Traffic: 1971-80

The problem remains, though, that too much of BR's freight business is linked to heavy industry with all the associated fluctuations in its fortunes.

The return of Conservative administrations in the General Elections of both 1979 and 1983 placed the future of Britain's railways in a highly uncertain position. Whereas post-war Labour governments had maintained a consistent commitment to rail transport, which

involved extensive public subsidy after 1968, the governments of Margaret Thatcher sought movement towards a more commercial footing as part of their plan to regulate the size of the Public Sector Borrowing Requirement. Investment criteria were also tightened and this invariably slowed the potential pace of improvement. Indeed some parts of the rail system have been forced into a 'knife-edge' existence, with run-down stock and equipment and barely adequate resources for basic maintenance, none of which provides a favourable user environment.

With sights on improved financial results and more efficient performance, a committee of enquiry was appointed under the chairmanship of Sir David Serpell. The resulting report of January 1983 outlined a number of network options for future railway operation. Option A presented an example of a commercially profitable network, totalling only 1,630 miles or sixteen per cent of the real network in 1983. At the opposite extreme was Option H involving a network of 10,070 route miles and a deficit of £803m. The Serpell report proved highly controversial and it remains to be seen whether any government would be bold enough to adopt any of its more radical alternatives. If any were to do so, the days of the railway in Britain would be well and truly over.

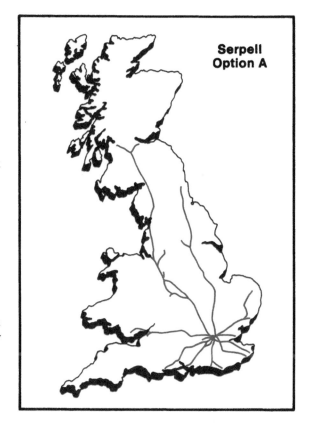

Plate 101: The former North British Railway's West Highland line at Spean Gorge — arguably the most scenic railway route in Britain but under persistent threat of closure.

NOW THEY ARE
CLOSED ... NOW
THEY ARE OPEN

**Major Preserved Railways
in Britain in 1980**

Plate 102: The spirit and atmosphere of another age is recaptured at Arley on the Severn Valley Railway, as Ivatt class 2MT 2-6-0 no. 46521 sets off for Bridgnorth.

Foreigners have often remarked upon the Englishman's passion for the steam locomotive. The map opposite bears eloquent testimony to this. Since the Beeching 'axe' and the demise of steam operation on the nationalised system, the desire to recapture the aura, vitality and charm of the British steam railway has gathered rapid momentum. Preservation societies have been formed and operating companies founded with the express purpose of re-creating parts of a world which would otherwise be lost to living memory. The task has been no easy one. That it has succeeded as far as it has is due to the dedication of those involved. It is also an indication of the growing importance of the 'leisure economy'. The lines are spread widely through the country, witness to the grass-roots foundation from which the preservation movement draws.

Plate 103: SR U class 2-6-0 no. 31806 exemplifies the Mid Hants Railway's high standards of restoration. It was photographed at the head of a Press special in the attractive LSWR setting of Alresford station.

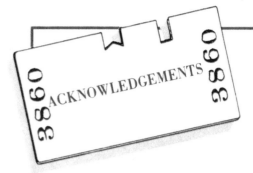

ACKNOWLEDGEMENTS

The authors record their thanks to the many people who have given advice and assistance in the preparation of this volume. Special mention must go to Miss Elspeth Buxton, Librarian to the Oxford University School of Geography, for her help in the task of assembling data for the maps and diagrams. Many of the photographs have benefited from the able attention of Tony Lee, photographic technician to the Oxford School of Geography.

The cartographic work has been done exclusively by Mrs Jayne Lewin and the authors record here their appreciation of the patient way in which rough drafts were transformed into highly presentable maps and diagrams.

Sources

Unless otherwise credited, the data employed in this atlas are drawn from official publications, notably *Parliamentary Papers*. Guides to further reading are George Ottley's *Bibliography of British Railway History* (HMSO 1983) and the *Journal of Transport History* (Leicester University Press to 1979, Manchester University Press from 1980).

Photographs are reproduced by courtesy of the following:

G.S. Ayers: 42, 43, 58, 61, 65, 67

Bodleian Library, Oxford: 8, 18 (G.A. Engl. Railways A.1); 9, 15 (G.A. fol. C.4); 10 (N.2706 d. 10/9. 1845. p. 47; 11 (N.2706 d. 10/8. 1845. p. 149)

British Rail: 32, 79, 80, 82, 88, 89, 90, 95, 98

European Railways Collection: 4, 50, 53, 54, 77, 97, 99, 102, 103

W.M.J. Jackson: 81

National Railway Museum: 3, 13, 16, 17, 19, 21, 22, 23, 24, 25, 26, 27, 28, 29, 30, 31, 33, 34, 35, 36, 37, 38, 39, 40, 41, 44, 45, 46, 47, 48, 49, 51, 52, 55, 68, 69, 70, 71, 73, 74, 76, 78, 94, 101

National Railway Museum/Box Collection: 66

National Railway Museum/Carrier Collection: 62

National Railway Museum/Cawston Collection: 56, 59, 60, 63, 83, 85, 87

National Railway Museum/Russell-Smith Collection: 84, 86, 91

National Railway Museum/Soole Collection: 64, 72, 75

National Railway Museum/Stephen Collection: 57

National Railway Museum/Williams Collection: 92, 93, 96, 100

University of Oxford School of Geography: 1, 2, 5

Photographs not credited above are in the possession of the authors.

For Product Safety Concerns and Information please contact our EU
representative GPSR@taylorandfrancis.com Taylor & Francis Verlag GmbH,
Kaufingerstraße 24, 80331 München, Germany

Printed and bound by CPI Group (UK) Ltd, Croydon, CR0 4YY

01/05/2025

01858550-0001